MW00388276

High Voltage Circuit Breakers

ELECTRICAL ENGINEERING AND ELECTRONICS

A Series of Reference Books and Textbooks

EXECUTIVE EDITORS

1. Rational Fault Analysis, *edited by Richard Saeks and S. R. Liberty*
2. Nonparametric Methods in Communications, *edited by P. Papantoni-Kazakos and Dimitri Kazakos*
3. Interactive Pattern Recognition, *Yi-tzuu Chien*
4. Solid-State Electronics, *Lawrence E. Murr*
5. Electronic, Magnetic, and Thermal Properties of Solid Materials, *Klaus Schröder*
6. Magnetic-Bubble Memory Technology, *Hsu Chang*
7. Transformer and Inductor Design Handbook, *Colonel Wm. T. McLyman*
8. Electromagnetics: Classical and Modern Theory and Applications, *Samuel Seely and Alexander D. Poularikas*
9. One-Dimensional Digital Signal Processing, *Chi-Tsong Chen*
10. Interconnected Dynamical Systems, *Raymond A. DeCarlo and Richard Saeks*
11. Modern Digital Control Systems, *Raymond G. Jacquot*

Additional Volumes in Preparation

High Voltage Circuit Breakers
Design and Applications

Ruben D. Garzon
Square D Co., Smyrna, Tennessee

Marcel Dekker, Inc.   New York · Basel

Library of Congress Cataloging–in–Publication Data

Garzon, R. D. (Ruben D.)
    High voltage circuit breakers : design and applications / Ruben Garzon.
        p.   cm. — (Electrical engineering and electronics : v. 100)
    Includes index.
    ISBN 0-8247-9821-X (hardcover : alk. paper)
        1. Electric circuit-breakers. 2. Electric power distribution-
    -High tension—Equipment and supplies. I. Title. II. Series.
    TK2842.G27   1996
    621.31'7—dc21
                                                                            96-37206
                                                                                CIP

The publisher offers discounts on this book when ordered in bulk quantities. For more information, write to Special Sales/Professional Marketing at the address below.

This book is printed on acid-free paper.

MARCEL DEKKER, INC.
270 Madison Avenue, New York, New York 10016

Current printing (last digit):
10 9 8 7 6 5 4 3 2

PRINTED IN THE UNITED STATES OF AMERICA

To my wife, Maggi, and to
Gigi, Mitzi and Natalie

The four pillars of my life

PREFACE

Ever since the time when electrical energy was beginning to be utilized, there has been a need for suitable switching devices capable of initiating and interrupting the flow of the electric current. The early designs of such switching devices were relatively crude and the principles of their operation relied only on empirical knowledge. Circuit breakers were developed on the basis of a "cut and try" approach, but as the electrical system capacity continued to develop and grow, a more scientific approach was needed to achieve optimized designs of circuit breakers that would offer higher performance capabilities and greater reliability.

The transition of current interruption from being an empirical art to an applied science began in the 1920's. It was only then that worldwide research started to unravel the subtleties of the electric arc and its significance on the current interruption process. Since those early research days, there has been a great deal of literature on the subject of current interruption. There have also been numerous technical articles on specific applications of circuit breakers, but most of these publications are highly theoretical. What has been missing are publications geared specifically to the needs of the practicing engineer—a simple source of reference that provides simple answers to the most often asked questions: Where does this come from? What does it mean? What can I do with it? How can I use it? How can I specify the right kind of equipment?

Circuit breakers are truly unique devices. They are a purely mechanical apparatus connected to the electrical system. They must systematically interact with the system, providing a suitable path for the flow of the electric current; furthermore, they must provide protection and control of the electric circuit by either initiating or stopping the current flow. Combining these tasks into one device requires a close interaction of two engineering disciplines. A good understanding of mechanical and electrical engineering principles is paramount for the proper design and application of any circuit breaker.

The purpose of this book is to bridge the gap between theory and practice, and to do so without losing sight of the physics of the interruption phenomena. The approach is to describe the most common application and design requirements and their solutions based on experience and present established practices. A strictly mathematical approach is avoided; however, the fundamentals of the processes are detailed and explained from a qualitative point of view.

Beginning with a simplified qualitative, rather than quantitative, description of the electric arc and its behavior during the time when current is being interrupted, we will then proceed to describe the response of the electric system and the inevitable interaction of current and voltage during the critical initial microseconds following the interruption of the current. We will show the specific behavior of different types of circuit breakers under different conditions.

After explaining what a circuit breaker must do, we will proceed to describe the most significant design parameters of such device. Particular emphasis will be placed in describing the contacts, their limitations in terms of continuous current requirements and possible overload conditions, and their behavior as the result of the electromagnetic forces that are present during short circuit conditions and high inrush current periods. Typical operating mechanisms will be described and the terminology and requirements for these mechanisms will be presented.

Over the years performance standards have been developed not only in the United States but in other parts of the world. Today, with the world tending to become a single market, it is necessary to understand the basic differences among these standards. Such an understanding will benefit anyone who is involved in the evaluation of circuit breakers designed and tested according to these different standards.

The two most widely and commonly recognized standards documents today are issued by the American National Standards Institute (ANSI) and the International Electrotechnical Commission (IEC). The standards set fort by these two organizations will be examined and their differences will be explained. By realizing that the principles upon which they are based are mainly localized operating practices, it is hoped that the meaning of each of the required capabilities will be thoroughly understood. This understanding will give more flexibility to the application engineer for choosing the proper equipment for any specific application and to the design engineer for selecting the appropriate parameters upon which to base the design of a circuit breaker, which must meet the requirements of all of the most significant applicable standards if it is to be considered of world-class.

This type of book is long overdue. For those of us who are involved in the design of these devices, it has been a long road of learning. Many times, not having a concise source of readily available collection of design tips and general design information, we have had to learn the subtleties of these designs by experience. For those whose concern is the application and selection of the devices, there is a need for some guidance that is independent of commercial interests. There have been a number of publications on the subject, but most, if not all of them, use a textbook approach to treat the subject with a strict mathematical derivation of formulae. The material presented here is limited to what is believed to be the bare essentials, the fundamentals of the fundamentals, and the basic answers to the most common questions on the subject.

The book *Circuit Interruption: Theory and Techniques*, edited by Thomas Browne, Jr. (Marcel Dekker, 1984) partially meets the aims of the new publication, but earlier works have become obsolete since some of the new design concepts of interrupter designs and revisions to the governing standards were not thoroughly covered. None of these previous publications covers specific design details, application, interpretation of standards, and equipment selection and specification.

There are a great number of practicing electrical engineers in the electrical industry, whether in manufacturing, industrial plants, construction, or public utilities, who will welcome this book as an invaluable tool to be used in their day-to-day activities.

The most important contributors to this book are those pioneer researchers who laid the foundations for the development of the circuit breaker technology. I am specially indebted to Lorne McConnell, from whom I learn the trade and who encouraged me in my early years. I am also indebted to all those who actively helped me with their timely comments, and especially to the Square "D" Co. for its support on this project. Most of all, I am especially grateful to my wife, Maggi, for her support and patience during the preparation of this book.

Ruben D. Garzon

CONTENTS

High Voltage Circuit Breakers

1

THE FUNDAMENTALS OF ELECTRIC ARCS

1.0 Introduction

From the time when the existence of the electric current flow was first established and even before the basic thermal, mechanical, and chemical effects produced by such current were determined, it had became clear that there was a need for inventing a device capable of initiating and of stopping the flow of the current.

Fundamentally, there are two ways in which the flow of current can be stopped, one is to reduce the driving potential to zero, and the other is to physically separate or create an open gap between the conductor that is carrying the current. Historically, it has been the later method the one most commonly used to achieve current interruption.

Oersted, Ampere and Faraday are among the first known users of circuit breakers and according to recorded history those early circuit breakers are known to have been a mercury switch which simply consisted of a set of conducting rods that were immersed in a pool of mercury.

Later in the evolutionary history of the current switching technology, the mercury switch, or circuit breaker was replaced by the knife switch design, which even now is still widely used for some basic low voltage, low power applications. Today, under the present state of the art in the current interruption technology, the interruption process begins at the very instant when a pair of electric contacts separate. It continues as the contacts recede from each other and as the newly created gap is bridged by a plasma. The interruption process is then finally completed when the conducting plasma is deprived of its conductivity.

By recognizing that the conducting plasma is nothing more than the core of an electric arc, it becomes quite evident that inherently the electric arc constitutes a basic, indispensable, and active element in the process of current interruption. Based on this simple knowledge, it follows that the process of extinguishing the electric arc constitutes the foundation upon which current interruption is predicated. It is rather obvious then that a reasonable knowledge of the fundamentals of arc theory is essential to the proper understanding of the interrupting process. It is intended that the following basic review describing the phenomena of electric discharges, will serve to establish the foundation for the work that is to be presented later dealing with current interruption.

1

1.1 Basic Theory of Electrical Discharges.

The principles that govern the conduction of electricity through, either a gas, or a metal vapor, are based on the fact that such vapors always contain positive and negative charge carriers, and that all types of discharges always involve the very fundamental processes of production, movement, and final absorption of the charge carriers as the means of conveying the electric current between the electrodes.

For the sake of convenience and in order to facilitate the review of the gas discharge phenomena, the subject will be divided into the following three very broad categories:

a) The non-self-sustaining discharge
b) The self-sustaining discharge and
c) The electric arc

1.1.1 Non-Self-Sustaining Discharges

When a voltage is applied across two electrodes, the charge carriers are acted upon by a force that is proportional to the electric field strength, this force establishes a motion of the ions towards the cathode and of the electrons towards the anode. When the moving charges strike the electrodes they give up their charges; thus producing an electric current through the gaseous medium. A continuous flow of current can take place only if the carriers whose charges are absorbed by the electrodes are continuously replaced. The replacement of the charge carriers can be made by a number of ionization processes such as, photoelectric, or thermionic emissions.

Initially, the discharge current is very small; however as the voltage is increased it is observed that the current increases in direct proportion to the voltage applied across the electrodes, until a level is reached where the charge carriers are taken by the electrodes at the same rate as they are produced. Once this equilibrium state is attained the current reaches a first recognizable stable limit that is identified as the saturation current limit. The value of the saturation current is dependent upon the intensity of the ionization, it is also proportional to the volume of gas filling the space between the electrodes and to the gas pressure.

At the saturation limit the current remains constant despite increases of the supply voltage to levels that are several times the level originally required to reach the saturation current limit. Because the saturation current is entirely dependent on the presence of charge carriers that are supplied by external ionizing agents this type of discharge is called a non-self-sustaining discharge.

Since the charge carriers are acted upon not only by the force exerted by the electric field, but by electro-static forces that are due to the opposite polarity of the electrodes, the originally uniform distribution of the charge carriers can be altered by the application of a voltage across the electrodes. It can be observed that an increase in the electrode's potential produces an increased concentration of electrons near the anode and of positive molecular ions near the cathode; thus creating what is known as space charges at the electrode boundaries.

The space charges lead to an increase in the electric field at the electrodes, which will result in a decrease of the field in the space between the electrodes. The drops of potential at the electrodes are known as the anode fall of potential, or anode drop, and as the cathode fall of potential, or cathode drop.

It was mentioned previously, that whenever the current reaches its saturation value the voltage applied across the electrodes and hence the electric field, may be substantially increased without causing any noticeable increase in the discharge current. However as the electric field strength increases, so does the velocity of the charge carriers, and since an increase in velocity represents an increase in kinetic energy it is logical to expect that when these accelerated charges collide with neutral particles new electrons will be expelled from these particles; thus, creating the condition known as shock ionization.

In the event that the kinetic energy is not sufficient for fully ionizing a particle it is possible that it will be enough to re-arrange the original grouping of the electrons by moving them from their normal orbits to orbits situated at a greater distance from the atom nucleus. This state is described as the excited condition of the atoms; once this conditions is reached a smaller amount of energy will be required in order to expel the shifted electron from this excited atom and to produce complete ionization. It is apparent then, that with a lesser energy level of the ionizing agent, successive impacts can initiate the process of shock ionization.

The current in the region of the non-self-sustaining discharge ceases as soon as the external source is removed; However when the voltage reaches a certain critical level the current increases very rapidly and a spark results in the establishing of a self-sustained discharge in the form of either a glow discharge or the electric arc.

In many cases, such as for example between parallel plane electrodes, the transition from a non-self sustaining to a self-sustaining discharge leads to an immediate complete puncture or flashover which, provided that the voltage source is sufficiently high, will result in a continuously burning arc being established. In the event that a capacitor is discharged across the electrodes, the resulting discharge takes the form of a momentary spark.

In other cases, where the electric field strength decreases rapidly as the distance between the electrodes increases, the discharge takes the form of a partial flashover. In this case the dielectric strength of the gas space is exceeded only near the electrodes and as a result a luminous discharge known as "corona" appears around the electrodes.

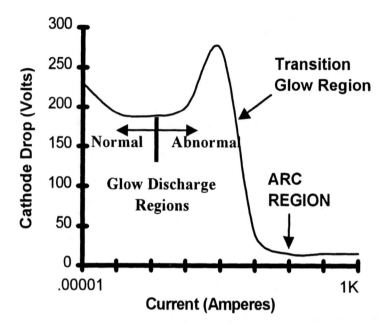

Figure 1.1 Schematic representation of the voltage-current relationship of a self-sustaining electrical discharge.

1.1.2 Self Sustaining Discharges

The transition from a non-self-sustaining discharge to a self-sustained discharge is characterized by an increase in the current passing though the gas, whereas the voltage across the electrodes remains almost constant. When the electrode potential is increased to the point where ionization occurs freely, the positive ions produced in the gas may strike the cathode with a force that is sufficient to eject the number of electrons necessary for maintaining the discharge. Under these circumstances no external means of excitation are needed and the discharge is said to be self-sustaining.

During the initial stages of the self-sustaining discharge the current density is only in the order of a few micro-amperes per square centimeter, the discharge has not yet become luminous and consequently it is called a dark discharge; However as the current continues to increase a luminous glow appears across the gas region between the electrodes, and, as illustrated in figure 1.1, the stage known as the "glow discharge" takes place and the luminous glow then becomes visible. The colors of the glow differ between the various glowing regions and vary in accordance with the surrounding gas. In air, for example, the negative or cathode glow exhibits a very light bluish color and the positive column is salmon pink. The glow discharge characteristic have proven to be very important for applications dealing with illumination.

The region called the normal glow region is that in which the current is low and the cathode is not completely covered by the cathode glow, the cathode current density at this time is constant and is independent of the discharge current. When the current is increased so that the cathode is completely covered by the negative glow, the current density has increased and so has the cathode voltage drop and this region is called the abnormal glow region.

As the current increases in the abnormal glow region the cathode drop space decreases in thickness. This leads to a condition where the energy imparted to the positive ions is increased and the number of ionizing collisions encountered by an ion in the cathode drop space is decreased. The increased energy of the incoming positive ions increase the cathode temperature which in turn leads to a condition of thermionic emission, that subsequently results in an increase in current that is accompanied by a rapid collapse in the discharge voltage. During this transitional period, the physical characteristics of the discharge change from those of a glow discharge to those of a fully developed arc.

1.2 The Electric Arc

The electric arc is a self-sustained electrical discharge that exhibits a low voltage drop, that is capable of sustaining large currents, and that it behaves like a non-linear resistor. Though the most commonly observed arc discharge occurs across air at atmospheric conditions, the arc discharge is also observed at high and low pressures, in a vacuum environment, and in a variety of gases and metal vapors. The gases and vapors, that serve as conductors for the arc, originate, partly from the electrodes, and partly from the surrounding environment and from their reaction products. The description of the electric arc will be arbitrarily divided into two separately identifiable types of arcs. This is done only as an attempt to provide with a simpler way to relate to future subjects dealing with specific interrupting technologies. The first arc type will be identified as the high pressure arc and the second type, which is an electric arc burning in a vacuum environment, will be identified as a low pressure arc.

1.2.1 High Pressure Arcs

High pressure arcs are considered to be those arcs that exist at, or above atmospheric pressures. The high pressure arc appears as a bright column characterized by having a small highly visible, brightly burning core consisting of ionized gases that convey the electric current. The core of the arc always exist at a very high temperature and therefore the gases are largely dissociated. The temperature of the arc core under conditions of natural air cooling reach temperatures of about 6000° K while when subjected to forced cooling, temperatures in excess of 20000° K have been observed. The higher temperatures that have been observed when the arc is being cooled at firsts appears to be a contradiction. One would think that under forced cooling conditions the temperature should be lower, however the higher temperature is the result of a reduction in the arc diameter which produces an increase in the current density of the plasma and consequently leads to the observed temperature increase.

By comparing the cathode region of the arc with the cathode region of the glow discharge, it is seen that the cathode of the glow discharge has a fall of potential in the range of 100 to 400 volts, it has a low current density, the thermal effects do not contribute to the characteristics of the cathode and the light emitted from the region near the cathode has the spectrum of the gas surrounding the discharge. In contrast the cathode of the arc has a fall of potential of only about 10 volts, a very high current density, and the light that is emitted by the arc has the spectrum of the vapor of the cathode material.

Some of the most notable arc characteristics, that have a favorable influence during the interrupting process, are the fact that the arc can be easily influenced and diverted by the action of a magnetic field or by the action of a high pressure fluid flow and that the arc behaves as a non-linear ohmic resistance. If the arc behaves like a resistors it follows then that the energy absorbed in the arc is equal to the product of the arc voltage drop and the current flowing through the arc.

Under constant current conditions the steady state arc is in thermal equilibrium, which means that the power losses from the arc column are balanced by the power input into the arc. However due to the energy storage capability of the arc there is a time lag between the instantaneous power loss and the steady state losses and therefore at any given instant the power input to the arc, plus the power stored in the arc is equal to the power loss from the arc. This time lag condition, as it will be seen later in this chapter, is extremely significant during the time of interruption, near current zero.

As a result of the local thermal equilibrium it is possible to treat the conducting column of the arc as a hot gas which satisfies the equations of conservation of mass, momentum and energy. To which, all the thermodynamic laws and Maxwell's electromagnetic equations apply. This implies that the gas composition, its thermal, and its electrical conductivity are factors which are essentially temperature determined.

The voltage drop across an arc can be divided into three distinct regions, as illustrated in figure 1.2. For short arcs the voltage drop appearing in a relatively thin region immediately in front of the cathode, represents a rather large percentage of the total arc voltage. This voltage drop across the region near the

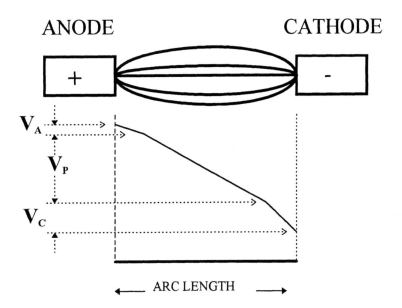

Figure 1.2 Voltage distribution of an arc column, V_A, represents the anode voltage; V_C, represents the cathode voltage, and V_P, represents the positive column voltage.

cathode is typically between 10 to 25 volts and is primarily a function of the cathode material. In the opposite electrode, the anode drop, is generally between 5 to 10 volts. The voltage drop across the positive column of the arc is characterized by a uniform longitudinal voltage gradient, whose magnitude, in the case of an arc surrounded by a gaseous environment, depends primarily on the type of gas, the gas pressure, the magnitude of the arc current, and the length of the column itself. Voltage values for the positive column gradient in the range of only a few volts per centimeter to several hundred volts per centimeter have been observed.

   The first extensive study of the electric arc voltage relationships, for moderate levels of current and voltage, was made by Hertha Ayrton [1], who developed a formula defining the arc voltage on the basis of empirical experimental results. The relationship still is considered to be valid, and is still widely used, although within a limited range of current and voltage.

   The classical Ayrton equation is given as:

$$e_0 = A + Bd + \frac{C + Dd}{i}$$

where:

e_0 = arc voltage
d = arc length
i = arc current
A=19, B=11.4, C=21.4 and D=3

The values of these constants A, B, C, and D, are empirical values, for copper electrodes in air.

The current density at the cathode is practically independent of the arc current, but it is strongly dependent upon the electrode material. In refractory materials such as carbon, tungsten, or molybdenum, that have a high boiling point, the cathode spot is observed to be relatively fixed, the cathode operates by thermionic emission and its current density is in the order of 10^3 amps per cm^2. The "cold cathode arc" is characteristic of low boiling point materials such as copper and mercury. The cathode spot in these materials is highly mobile, it operates in some form of field emission and its current density is in the order of 10^6 to 10^7 amps per cm^2. In those materials that have a low boiling point a considerable amount of material is melted away from the electrodes; while the material losses of refractory materials is only due to vaporization. Under identical arcing conditions the refractory material losses are considerably less than the losses of low boiling point materials, and consequently this constitutes an important factor that must be kept on mind when selecting materials for circuit breaker contacts.

1.2.2 Low Pressure (Vacuum) Arcs

The low pressure or vacuum arc, like those arcs that occur at, or above atmospheric pressure, share most of the same basic characteristics just described for the electric arc, but the most significant differences are: (a) An average arc voltage of only about 40 volt which is significantly lower than the arc voltages observed in high pressure arcs. (b) The positive column of the vacuum arc is solely influenced by the electrode material because the positive column is composed of metal vapors that have been boiled off from the electrodes; while in the high pressure arc the positive column is made up of ionized gases from the arc's surrounding ambient. (c) Finally, and, perhaps the most significant and fundamental difference which is the unique characteristic of a vacuum arc that allows the arc to exist in either a diffuse mode or in a coalescent or constricted mode.

The diffuse mode is characterized by a multitude of fast moving cathode spots and by what looks like a multiple number of arcs in parallel. It should be pointed out that this is the only time when arcs in parallel can exist without the need of balancing, or establishing inductances. The magnitude of the current being carried by each of the cathode spots is a function of the contact material, and in most cases is only approximately 100 amperes. Higher current densities

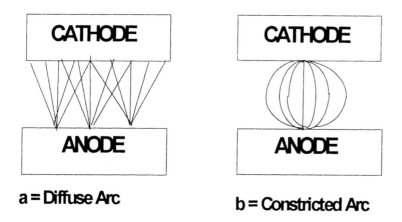

a = Diffuse Arc

b = Constricted Arc

Figure 1.3 Outline of an arc in vacuum illustrating the characteristics of the arc in a diffuse and in a constricted mode.

are observed on refractory materials such as tungsten or graphite, while lower currents correspond to materials that have a low boiling point such as copper.

When the current is increased beyond a certain limit, that depends on the contact material, one of the roots of the arc gets concentrated into a single spot at the anode while the cathode spots split to form a closely knit group of highly mobile spots as shown in figure 1.3. If the cathode spots are not influenced by external magnetic fields, they move randomly around the entire contact surface at very high speeds.

When the current is increased even further, a single spot appears at the electrodes. The emergence of a single anode spot is attributed to the fact that large currents greatly increase the collision energy of the electrons and consequently when they collide with the anode, metal atoms are released thus producing a gross melting condition of the anode. There is a current threshold at which the transition from a diffuse arc to a constricted arc mode takes place, this threshold level is primarily dependent upon the electrode size and the electrode material. With today's typical, commercially available vacuum interrupters, diffused arcs generally occur at current values below 15 kilo-amperes and therefore in some ac circuit breaker applications it is possible for the arc to change from a diffuse mode to a constricted mode as the current approaches its peak and then back to a diffuse mode as the current approaches its natural zero crossing. It follows then, that the longer the time prior to current zero that an interrupter is in the diffuse mode, the greater it is its interrupting capability.

1.3 The Alternating Current Arc

The choice of sinusoidal alternating currents as the standard for the power systems is indeed a convenient and fortuitous one in more ways than one. As it was described earlier, in the case of an stable arc, as the arc current increases the arc resistance decreases due to the increase in temperature which enhances the ionizing process, and when the current decreases the ionization level also decreases while the arc resistance increases; thus, there is a collapse of the arc shortly before the alternating current reaches its normal zero value at the end of each half cycle. The arc will reignite again when the current flows in the opposite direction, during the subsequent half cycle, provided that the conditions across the electrodes are still propitious for the existence of the arc. The transition time between the two half cycles is greatly influenced by the medium on which the arc is being produced and by the characteristics of the external circuit.

The arc current as it approaches zero is slightly distorted from a true sine wave form due to the influence of the arc voltage, and therefore the arc is extinguished just prior to the nominal current zero crossing. The current zero transition is accompanied by a sharp increase in the arc voltage, and the peak of this voltage is defined as the peak of extinction voltage. When the peak of the extinction voltage reaches a value equal to the instantaneous value of the voltage applied to the arc by the circuit, the arc current can not be maintained any longer and thereafter, the current in the opposite direction cannot be reestablished immediately; thus, at every current zero there is a finite time period where there can not be any current flow. This is the time period generally referred as the "current zero pause". During the zero current period, the discharge path is partially de-ionized on account of the heat losses and therefore the electric field needed to re-establish the arc after the reversal of the current becomes greater than the field required to maintain the arc. This means that the required reignition voltage is higher than the voltage which is necessary to sustain the arc, and therefore the current will remain at its zero value until the reignition voltage level is reached. If the arc is re-established the current increases and the voltage falls reaching its minimum value, which is practically constant during most of the half cycle, and in the region of maximum current.

The sequence that have just been described will continue to repeat itself during each one of the subsequent half cycles, provided that the electrodes are symmetrical; however in most cases there will be some deviations in the arc's behavior which arise from differences in the electrode materials, cooling properties, gas ambient, etc. This asymmetric condition is specially accentuated when the electrodes are each made of different materials.

The time window that follows a current zero and during which arc reignition can occur depends upon the speed at which the driving voltage increases at the initiation of each half cycle, and on the rate at which de-ionization takes place in the gap space; in other words, the reignition process represents the relationship between the rate of recovery of the supply voltage and the rate of de-ionization, or dielectric recovery of the space across the electrode gap.

1.4 The Current Interruption Process

In the preceding paragraphs the electric arcs were assumed to be either static, as in the case of direct current arcs, or quasi static, as in the case of alternating current arcs. What this means is that we had assumed that the arcs were a sustained discharge, and that they were burning continuously. However, what we are interested in, is not with the continuously burning arc; but rather with those electric arcs which are in the process of being extinguished because, as we have learned, current interruption is synonymous with arc extinction, and since we further know that the interrupting process is influenced by the characteristics of the system and by the arc's capacity for heat storage we can expect that the actual interruption process, from the time of the initial creation of the arc until its extinction, will depend primarily on whether the current changes in the circuit are forced by the arc discharge, or whether those changes are controlled by the properties of the power supply. The first alternative is what it can be observed during direct current interruption where the current to be interrupted must be forced to zero. The second case is that of the alternating current interruption where a current zero occurs naturally twice during each cycle.

1.4.1 Interruption of Direct Current

Although the subject relating to direct current interruption does not enter into the later topics of this book, a brief discussion explaining the basics of direct current interruption is given for reference and general information purposes.

The interruption of direct current sources differ in several respects from the phenomena involving the interruption of alternating currents. The most significant difference is the obvious fact that in direct current circuits there are no natural current zeroes and consequently a current zero must be forced in some fashion in order to achieve a successful current interruption. The forcing of a current zero is done either by increasing the arc voltage to a level that is equal, or higher than the system voltage, or by injecting into the circuit a voltage that has an opposite polarity to that of the arc voltage, which in reality is the equivalent of forcing a reverse current flowing into the source.

Generally the methods used for increasing the arc voltage consist of simply elongating the arc column, of constricting the arc by increasing the pressure of the arc's surroundings so as to decrease the arc diameter and to increase the arc voltage, or of introducing a number of metallic plates, along the axis of the arc, in such a way so that a series of short arcs are developed.

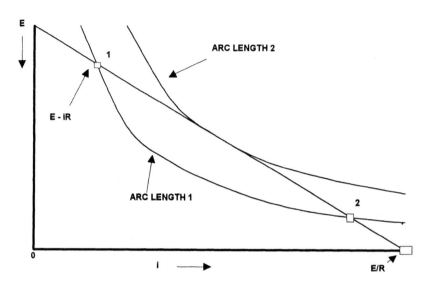

Figure 1.4 Relationships of arc and system voltages during the interruption of direct current.

With this latter approach it is easy to see that as a minimum the sum of the cathode and anode drops, which amounts to at least 30 volts, can be expected for each arc. Since the arcs are in series their voltage drop is additive, and the final value of the arc voltage is simply a function of the number of pairs of interposing plates. The second method, that is the driving of a reversed current, is usually done by discharging a capacitor across the arc. The former method is commonly used on low voltage applications, while the later method is used for high voltage systems.

For a better understanding of the role played by the arc voltage during the interruption of a direct current, let us consider a direct current circuit that has a voltage E, a resistance R, and an electric arc, all in series with each other. The current in the circuit, and therefore the current in the arc will adjust itself in accordance with the values of the source voltage E, the series resistance R, and the arc characteristics. By referring to figure 1.4, where the characteristics of the arc voltage are shown as a function of the current for two different arc lengths, it is seen that the arc voltage e_0 is smaller than the supply voltage E by an amount equal to iR, so that $e_0 = E - iR$. If the straight line that represents this voltage is plotted as shown in the figure, it is seen that this line intercepts with the curve representing the arc characteristics for length 1 at the points indicated as 1 and 2. Only at the intersection of these points, as dictated by their respective currents, it is possible to have an stable arc. If for example the current corresponding to point 2 increases it can be seen that the arc voltage is too low, and if the current decreases the corresponding arc voltage is too high, and therefore the current always will try to revert to its stable point. In order to obtain a stable condition at point 1 it will be necessary for the circuit to have a very high series resistance and a high supply voltage. We already know that the net result of lengthening the arc is an increase in the arc resistance and a reduction in current, provided that the supply voltage remains constant. However, and as it is generally the case in practical applications, some resistance is always included in the circuit, and the reduction of the current will produce a corresponding reduction in the voltage across the series resistance. The electrode voltage is thus increased until eventually, when the arc is extinguished, it becomes equal to the system voltage. These conditions are shown graphically in figure 1.4 for the limiting case where the arc characteristics of the longer arc represented by the curve labeled length 2 have reached a position where it no longer intercepts the circuit characteristics that are represented by the line E - iR, and therefore the condition where the arc can not longer exist has been reached.

When inductance is added to the circuit that we have just described we find that the fundamental equation for this inductive circuit can we represented as follows:

$$L\frac{di}{dt} = (E - iR) - e_a = \Delta e$$

This equation indicates that the inductive voltage produced during interruption is equal to the source voltage E reduced by the voltage drop across the inherent resistance of the circuit, and by the arc voltage. For the arc to be extinguished the current i must continually decrease which in turn suggests that the derivative of the current (di/dt) must be negative.

As it is seen from the equation, the arcing conditions at the time of interruption are significantly changed, in relation to the magnitude of the inductance L. Since the inductance opposes the current changes, the falling of the current results in an induced e.m.f. which acts additively to the source voltage. The relationship between the source voltage and the arc voltage still holds for the inductive circuit, therefore it is necessary to develop higher arc voltages which require that the rupturing arc length be increased to provide the additional voltage.

It is also important to remember that when interrupting a direct current circuit, the interrupting device must be able to dissipate the total energy that is stored in the circuit inductance.

1.4.2 Interruption of Alternating Current

As it has been shown in the previous section, in order to extinguish a direct current arc it is required to create, or in some way, to force a current zero. In an alternating current circuit the instantaneous value of the current passes through zero twice during each cycle and therefore the zero current condition is already self-fulfilled; consequently, to interrupt an alternating current it is only sufficient to prevent the reignition of the arc after the current has passed through zero. It is for this reason that de-ionization of the arc gap close to the time of a natural current zero is of the utmost importance. While any reduction of the ionization of the arc gap close to the point of a current peak it is somewhat beneficial, this action does not significantly aid in the interrupting process. However, because of thermodynamic constrains that exist in some types of interrupting devices, it is advisable that all appropriate measures to enhance interruption be taken well in advance of the next natural current zero at which interruption is expected to take place. Successful current interruption depends on whether the dielectric withstand capability of the arc gap is greater than the increasing voltage that is being impressed across the gap by the circuit in an attempt to re-establish the flow of current. The dielectric strength of the arc gap is primarily a function of the interrupting device; while the voltage appearing across the gap is a function of the circuit constants.

At very low frequencies, in the range of a few cycles but well below the value commonly used in general for power frequencies, the rate of change of the current passing through zero is very small, and in spite of the heat capacity of the arc column, the temperature and the diameter of the arc have sufficient time to adjust to the instantaneous values of the current and therefore when the current drops to a sufficiently small value, (less than a few amperes, depending upon the contact gap), the alternating current arc will self extinguish, unless the voltage at the gap contacts at the time of interruption is sufficiently high to produce a glow discharge.

Normal power application frequencies, which are generally in the range of 16 2/3 to 60 Hertz, are not sufficiently low to ensure that the arc will go out on its own. Experience has shown that an alternating current arc, supported by a 50 Hz. system of 30 kilovolts, that is burning across a pair of contacts in open air and up to 1 meter in length, can not be extinguished. Special measures need to be taken if the effective current exceeds about 10 amperes. This is due to the fact that at these frequencies when the current reaches its peak value the electric conductivity of the arc is relatively high and since the current zero period is very short the conductivity of the arc, if the current is relatively large, can not be reduced enough to prevent re-ignition. However, since the current oscillates between a maximum positive and a maximum negative value there is a tendency to extinguish the arc at the current zero crossing due to the thermal lag

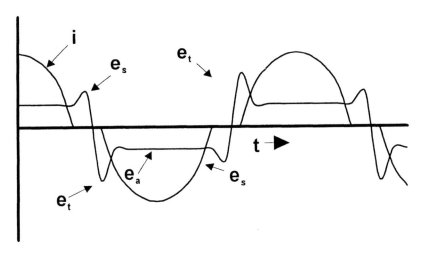

Figure 1.5 Typical variations of current and voltage showing the peak of extinction voltage e_s and peak of re-ignition voltage e_t.

previously mentioned. The time lag between temperature and current is commonly referred to as the "arc hysteresis".

When the alternating current crosses through its zero, the arc voltage takes a sudden jump to a value equal to the sum of the instantaneous peak value of the extinguishing voltage from the previous current loop, plus the peak value of the reignition voltage of the next current loop, which is associated with the reversal of the current. In the event that the arc is reignited, and immediately after the reignition has taken place, the arc voltage becomes relatively constant and of a significantly lower magnitude, as illustrated in figure 1.5, In order for a reignition to occur, the applied voltage must exceed the value of the total re-ignition voltage, (e_t). One practical application, derived from observing the characteristics of the extinction voltage is that during testing of an interrupting device, it provides a good indication of the behavior the device under test. A good, large and sharp peak of voltage indicates that the interrupter is performing adequately, but if the peak of the voltage begins to shown a smooth round top and the voltage magnitude begins to drop, it is a good indication that the interrupter is approaching its maximum interrupting limit. If reignition does not happen, the flow of the current will cease and therefore interruption will be accomplished. From what has been described, it is rather obvious that the most favorable conditions for interruption are those in which the applied voltage is at its lowest when the current reaches the zero value, however this ideal condition can be realized only with a purely resistive circuit.

1.4.2.1 Interruption of Resistive Circuits

In an alternating current circuit containing only resistance, or having a negligible amount of inductance, the current is practically in phase with the voltage, and during steady state operating conditions, both the current and the voltage reach their zero value simultaneously. But when a pair of contacts separate and an arc is developed between the contacts, the phase relationship, in theory it still exists, but in practice, as explained before the current will reach the zero value slightly ahead of the voltage. As the current (I) passes through zero the instantaneous value of the peak of extinction voltage, shown as e_s in figure 1.6, is equal to the instantaneous value of the applied voltage (E). From this point on no new charges are produced in the gas space between the contacts and those charges still present in the gas space are being neutralized by the de-ionized processes that are taking place. The gas space and the electrodes continue to increasingly cool down and therefore the minimum voltage required for the arc to re-ignite is increasing with time, the general idea of this increase is shown approximately in figure 1.6 in the curve marked as (1).

If the applied voltage E, shown as curve 2, rises at a higher rate than the re-ignition voltage e_t so that the corresponding curves intersect, the arc will be re-established and will continue to burn for an additional half cycle, at the end of which the process will be repeated. It will be assumed that during this time the gap length had increased and therefore the arc voltage and the peak of the extinction voltage can be assumed to be larger. The increase in the gap length will also provide an additional withstand capability and if the supply voltage is less than the re-ignition voltage then a successful interruption of the current in the circuit will be achieved.

1.4.2.2 Interruption of Inductive Circuits

Generally, in an inductive circuit the resistance is rather small in relation to the inductance and therefore there is a large phase angle difference between the voltage and the current. The current zero no longer occurs at the point where the voltage is approaching zero but instead when it is close to its maximum value. This implies that the conditions favor the re-striking of the arc immediately after the current reversal point.

It is important to note that in actual practice all inductive circuits have a certain small amount of self capacitance, such as that found between turns and coils in transformers and in the self-effective capacitance of the device itself in relation to ground. Although the effective capacitance, under normal conditions, can be assumed to be very small, it plays an important role during the interrupting process. The capacitance to ground appears as a parallel element to the arc, and therefore at the instant of current zero the capacitance is charged to a voltage equal to the maximum value of the supply voltage, plus the value of the peak of the extinction voltage.

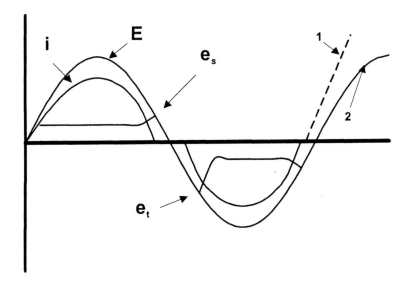

Figure 1.6 Interruption of a purely resistive circuit showing the current, voltages and recovery characteristics, for the electrode space (curve 1) and for the system voltage (curve 2).

When the arc is extinguished, the electro-magnetic energy stored in the inductance is converted into electrostatic energy in the capacitance and vice versa. The natural oscillations produced by the circuit are damped gradually by the effects of any resistance that may be present in the circuit, and since the oscillatory frequency of the inductance and the capacitance is much greater than the frequency of the source the supply voltage may be regarded as being constant during the time duration of the oscillatory response. These voltage conditions are represented in figure 1.7. During the interruption of inductive alternating circuits, the recovery voltage can be expected to reach its maximum value at the same time at which the current is interrupted, since the circuit is broken as the current approaches zero. However due to the inherent capacitance to ground the recovery voltage does not reach its peak at the same instant the current is interrupted and therefore, during this brief period, a transient response is observed in the circuit.

1.4.2.3 Interruption of Capacitive Circuits

The behavior of a purely capacitive circuit during the interruption process is illustrated in figure 1 8. It should be noted that, in contrast with the high de-

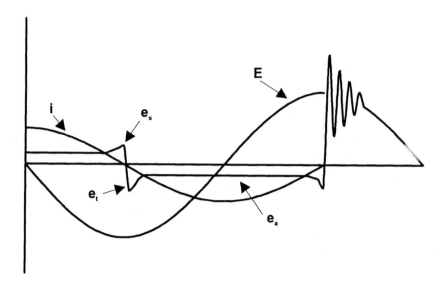

Figure 1.7 Current and voltage characteristics during interruption of an inductive circuit.

gree of difficulty that is encountered during the interruption of an inductive circuit, when interrupting a capacitive circuit, the system conditions are definitely quite favorable for effective interruption at the instant of current zero, because the supply voltage that appears across the electrodes is increased at a very slow rate. At the normal current zero where the arc interruption has taken place the capacitor is charged to, approximately, the maximum value of the system voltage. The small difference that may be observed is due to the arc voltage, how- ever the magnitude of the arc voltage is small in comparison to the supply voltage and generally it can safely be neglected.

At interruption, and in the absence of the current, the capacitor will retain its charge, and the voltage across the gap will be equal to the algebraic sum of the applied voltage and the voltage trapped in the capacitor. The total voltage increases slowly from an initial value equal to zero, until one half of a cycle later the voltage across the gap reaches twice the magnitude of the supply voltage; however there is a relative long recovery period that may enable the gap to recover its dielectric strength without reigniting. Under certain circumstances there may be restrikes that could lead to a voltage escalation condition. This particular condition will be discussed later in the section dealing with high voltage transients.

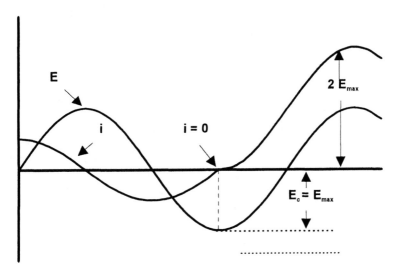

Figure 1.8 Voltage and current characteristics during the interruption of a capacitive circuit

1.5 Review of Main Theories of ac Interruption

The physical complexity in the behavior of an electric arc during the interrupting process, have always provided the incentive for researchers to develop suitable models that may describe this process. Over the years a variety of theories have been advanced by many researchers. In the early treatments of the interruption theory the investigative efforts were concentrated at the current zero region, which most obviously is the region when the alternating current arc either reignites or is extinguished. Recently, models of the arc near the current maximum have been devised for calculating the diameter of the arc. These models are needed since the arc diameter constitutes one of the critical dimensions needed for optimizing the geometry of the nozzles that are used in gas blasted interrupters. It should be recognized that all of these models provide only an approximate representation of the interrupting phenomena; but, as the research continues and with the aid of the digital computer, more advanced and more accurate models that include partial differential equations describing complex gas flow and thermodynamic relationships are being developed.

In the section that follows, a summary of qualitative descriptions of some of the early classical theories, and of some of the most notable recent ones is

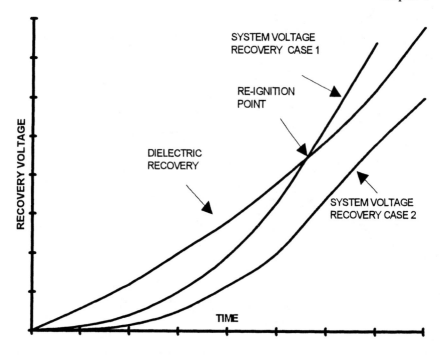

Figure 1.9 Graphical representation of the "Race Theory."

presented. The chosen theories are those that have proved to closely represent the physical phenomena, or to those that have been used as the basis for the development of more complex, combined modern day theories.

One of the earlier theories, usually referred to as the wedge theory, is now largely ignored, however it is briefly mentioned here because of the strong influence it had among researchers during the early research days in the area of arc interruption and of its application to the emerging circuit breaker technology.

1.5.1 Slepian's Theory

The first known formal theory of arc interruption was introduced by Joseph Slepian in 1928 [2]. The Slepian theory is also known as the "race theory it simply states that successful interruption is achieved whenever the rate at which the dielectric strength of the gap increases is faster than the rate at which the reapplied system voltage grows.

Slepian visualized the process of interruption as beginning immediately after a current zero when electrons are forced away from the cathode and when

a zone, or thin sheath composed of positive ions is created in the space immediately near the cathode region. He believed that the dielectric withstand of
this sheath had to be greater than the critical breakdown value of the medium
where interruption had taken place. The interrupting performance depended on
whether the rate of ion recombination, which results in an increase in the
sheath thickness, is greater than the rate of rise of the recovery voltage which
increases the electric field across the sheath.

The validity of this theory is still accepted, but within certain limitations.
The idea of the sheath effects is still important for predicting a dielectric
breakdown failure, which is the type of breakdown that occurs several hundreds of microseconds after current zero, when the ion densities are low. However this mechanism of failure is not quite as accurate for the case of thermal
failures, which generally occur at less than ten microseconds after current zero,
and when the ions densities are still significant, and when the sheath regions
are so thin that they can usually be neglected. The concept of the race theory is
graphically illustrated in figure 1.9.

1.5.2 Prince's Theory

This theory which is known as the displacement, or the wedge theory was ad-
vocated by Prince [3] in the U.S. and by Kesselring [4] in Germany. According
to this theory, the circuit is interrupted if the length of the gas discharge path
introduced into the arc increases during the interrupting period to such an
extent that the recovery voltage is not sufficiently high to produce a breakdown
in this path.

According to this theory as soon as the current zero period sets in, the arc is
cut in two by a blast of cooling gases and the partly conductive arc halves of
the arc column are connected in series with the column of cool gas which is
practically non-conductive. If it is assumed that the conductivity of the arc
stubs is high in comparison to that of the gas, then it can be assumed that the
stubs can be taken as an extension of the electrodes. In that case the dielectric
strength of the path between the electrodes is approximately equal to the
sparking voltage of a needle point gap in which breakdown is preceded by a
glow discharge. At the end of this current zero period, which corresponds to
the instant t_1 the two halves of the arc are separated by a distance

$$D = 2 v t$$

where v is the flow velocity of the cooling medium and t is the time duration
of the current period. Assuming, for example, the interruption of an air blast
circuit breaker where it is given the current zero time $t = 100$ microseconds
and the air flow velocity $v = 0.3$ millimeters per microsecond (which corre-

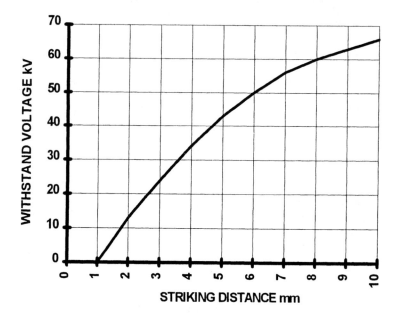

Figure 1.10 Withstand capability of a pair of plane electrodes in air at atmospheric pressure.

sponds to the velocity of sound in air), then using the previously given equation for the space of cool air, the distance between the electrodes is $D=60$ millimeters. Now referring to figure 1.10 we find that for a 60 mm distance the withstand capability should be approximately 50 kV.

1.5.3 Cassie's Theory

Among the first useful differential equation describing the dynamic behavior of an arc was the one presented by A. M. Cassie in 1939 [5]. Cassie developed his equation for the conductivity of the arc based on the assumption that a high current arc is governed mainly by convection losses during the high current time interval. Under this assumption a more or less constant temperature across the arc diameter is maintained. However, as the current changes so does the arc cross section, but not so the temperature inside of the arc column. These assumptions have been verified experimentally by measurements taken upstream of the vena contracta of nozzles commonly used in gas blast circuit breakers. Under the given assumptions, the steady state conductance G of the model is simply proportional to the current, so that the steady state voltage gradient E_0 is fixed. To account for the time lag that is due to the energy storage capacity

Q and the finite rate of energy losses N the concept of arc time constant θ is introduced. This "Time Constant" is given by:

$$\theta = \frac{Q}{N}$$

The following expression is a simplified form of the Cassie equation, this equation is given in terms of the instantaneous current.

$$\frac{d}{dt}(G^2) = \frac{2}{\theta}(\frac{I}{E_0})^2$$

For the high current region, data collected from experimental results is in good agreement with the model. However, around the current zero region, agreement is good only for high rates of current decay. Theoretically and practically, at current zero the arc diameter never decays to zero to result in arc interruption. At current zero there is a remaining small filament of an arc with a diameter of only a fraction of a millimeter. This filament still is a high temperature plasma that can be easily transformed into an arc by the reappearance of a high enough supply voltage. The Cassie model in many cases is referred as the high current region model of an arc. This model has proved to be a valuable tool for describing the current interruption phenomena specially when it is used in conjunction with another well known model, the Mayr model.

1.5.4 Mayr's Theory

Mayr took a radically different approach to that of Cassie, and in 1943 he published his theory [6]. He considered an arc column where the arc diameter is constant and where the arc temperature varies as a function of time, and of the radial dimension. He further assumed that the decay of the temperature of the arc was due to thermal conduction and that the electrical conductivity of the arc was dependent on temperature.

From an analysis of the thermal conduction in Nitrogen at 6000 Kelvin and below, Mayr found only a slow increase of heat loss rate in relation with the axial temperature; therefore he assumed a constant power loss N_0 which is independent of temperature or current. The resulting differential equation is:

$$\frac{1}{G}\frac{dG}{dt} = \frac{1}{\theta}(\frac{EI}{N_0} - 1)$$

where:

$$\theta = \frac{Q_0}{N_0}$$

The validity of this theory during the current zero period is generally acknowledged and most investigators have used the Mayr model near current zero successfully primarily because in this region the radial losses are the most dominant and controlling factor.

1.5.5 Browne's Combined Theory

It has been observed that in reality, the arc temperature is generally well above the 6000 K assumed by Mayr and it is more likely to be in excess of 20000 K. These temperatures are so high that they lead to a linear increase of the gas conductivity, instead of the assumed exponential relationship. To take into consideration these temperatures and in order to have a proper dynamic response representation it is necessary that, in a model like Mayr's, the model must follow closely an equation of the type of Cassie's during the current controlled regime.

T. E. Browne recognized this need and in 1948 [7] he developed a composite model using a Cassie like equation to define the current controlled arc regime and then converting to a Mayr like equation for the temperature controlled regime, and in the event that interruption does not occur at the intended current zero, he reverted again to the Cassie model.

The transition point where each of these equations are considered to be applicable is assumed to be at an instant just a few microseconds around the point where the current reaches its normal zero crossing.

In 1958 Browne extended the application of his combined model [8] to cover the analysis of thermal re-ignitions that occur during the first few microseconds following critical post current zero energy balance period.

Starting with the Cassie and the Mayr equations and assuming that before current zero the current is defined by the driving circuit, and that after current zero, the voltage applied across the gap is determined strictly by the arc circuit, Browne assumed that the Cassie equation is applicable to the high current region prior to current zero and also shortly after current zero following a thermal re-ignition. The Mayr equation was used as a bridge between the regions were the Cassie concept was applied. Browne reduced the Cassie and the Mayr equations to the following two expressions:

1). For the Cassie's period prior to current zero

$$\frac{d}{dt}\left(\frac{1}{R^2}\right) + \frac{2}{\theta}\left(\frac{1}{R^2}\right) = \frac{2}{\theta}\left(\frac{1}{E_0}\right)^2$$

2). for the Mayr's period around current zero

$$\frac{dR}{dt} - \frac{R}{\theta} = -\frac{e^2}{\theta N_0}$$

Experimental evidence [9] has demonstrated that this model is a valuable tool which has a practical application and which has been used extensively in the design and evaluation of circuit breakers. However, its usefulness depends on knowledge of the constant θ, which can only be deduced from experimental results. Browne calculated this constant from tests of gas blast interrupters and found it to be in the order of one microsecond, which is in reasonable agreement with the commonly accepted ranges found by other investigators [10].

1.5.6 Modern Theories

In recent years there has been a proliferation of mathematical models; However these models are mainly developments on more advanced methodologies for performing numerical analysis, using the concepts established by the classic theories that have been just described.

But also there has been a number of new more complex theories that have been proposed by several groups of investigators. Significant contributions have been made by: Lowke and Ludwig [11], Swanson [12], Frind [13], Tuma [14, 15], and Hermann, Ragaller, Kogelschatz, Niemayer, and Shade [16,17,18]. Among these works, probably the most significant technical contribution can be found in the investigations of Hermann and Ragaller [18].They developed a model that accurately describes the performance of air and of SF_6 interrupters. What is different in this model, from the earlier models, is that the effects of turbulence downstream of the throat of the nozzle are taken into consideration.

On this model the following assumptions are made:
a) There is a temperature profile which encompasses three regions; the first one embodies the arc core, the second that covers the arc's surrounding thermal layer and the third consisting of the external cold gas.
b) The arc column around current zero is cylindrical and the temperature distribution is independent of its axial position, and
c) The average gas flow velocities are proportional to the axial position.

The relatively consistent and close agreement, that has been obtained, between the experimental results and the theory suggests that, although some refinements may still be added, this is the model that so far has given the best description and the most accurate representation of the interruption process in a circuit breaker.

The models that have been listed in this section have as a common denominator the recognition of the important role played by turbulence in the interrupting process. Swanson [19] for example has shown that at 2000 A turbulence has a negligible effect on the arc temperature, while at 100 A turbulence makes a difference of 4000 Kelvin and at current zero the difference made by turbulence reaches values of over 6000 Kelvin. In a way these new models serve to probe or reinforce the validity of the Mayr equation since with the magnitudes of temperature reductions produced by the turbulent flow the arc column cools down to a range of values that are nearing those assumed by the Mayr equation where the electrical conductivity varies exponentially with respect to temperature.

REFERENCES

1. Ayrton H : The Electric Arc (D. Van Norstrand New York, 1902.
2. J. Slepian, Extinction of an a.c. Arc, Transactions AIEE 47 p. 1398, 1928.
3. D.C. Prince and W. F. Skeats, The oil blast circuit breaker, Transactions AIEE 50 pp 506-512, 1931.
4. F. Kesselring, Untersuchungen an elektrischen Lichtogen., ETZ. vol. 55, 92, 1932.
5. A. M. Cassie, Arc Rupture and circuit severity: A new theory, Internationale des Grands Reseaux Electriques'a Haute Tension (CIGRE), Paris, France, Report No. 102, 1939.
6. O. Mayr, Beitrage zur Theorie des Statisghen und des Dynamishen Lichtbogens, Archiv fur Electrotechnik, 37, 12, 588-608, 1943.
7. T. E. Browne, Jr. A study of arc behavior near current zero by means of mathematical models, AIEE Transactions 67: 141-143, 1948.
8. T. E. Browne Jr., An approach to mathematical analysis of a-c arc extiction in circuit breakers, AIEE Transactions 78 (Part III) : 1508-1517, 1959.
9. J. Urbanek, The time constant of high voltage circuit breaker arcs before current zero, Proc. IEEE 59: 502-508, April, 1971.
10. W. Reider, J. Urbanek, New Aspects of Current Zero Research on Circuit Breaker Reignition. A Theory of Thermal Non Equilibrium Conditions (CIGRE), Paper 107, 1966.
11. J.J. Lowke and H.C. Ludwig, A simple model for high current arcs stabilized by forced convection, Journal Applied Physics 46: 3352-3360, 1975.
12. B. W. Swanson and R. M. Roidt, Numerical solutions for an SF$_6$ arc, Proceedings IEEE 59: 493-501, 1971.
13. G. Frind and J.A Rich, Recovery Speed of an Axial Flow Gas Blast Interrupter, IEEE Transactions Pow. App. Syst. P.A.S.-93, 1675, 1974.
14. D.T. Tuma and J.J. Lowke, Prediction of Properties of arcs stabilized by forced convection, J. Appl. Phys. 46: 3361-3367, 1975.

15. D.T. Tuma and F. R. El-Akkari, Simulations of transient and zero current behavior of arcs stabilized by forced convection, IEEE Trans. Pow. Appar. Syst. PAS-96:1784-1788, 1977.

16. W. Hermann, U. Kogelschatz, L. Niemeyer, K. Ragaller, and E. Shade, Study of a high current arc in a supersonic nozzle flow, J. Phys. D. Appl. Phys. 7: 1703-1722, 1974.

17. W. Hermann, U. Kogelschatz, K. Ragaller, and E. Schade, Investigation of a cylindrical axially blown, high pressure arc, J. Phys. D: Appl. Phys. 7: 607, 1974

18. W. Hermann, and K. Ragaller, Theoretical description of the current interruption in gas blast breakers, Trans. IEEE Power Appar. Syst. PAS-96: 1546-1555, 1977.

19. B. W. Swanson, Theoretical models for the arc in the current zero regime, (edited by K. Ragaller) Plenum Press, New York: 137, 1978.

2

SHORT CIRCUIT CURRENTS

2.0 Introduction

During its operation an electric system is normally in a balanced, or a steady state condition. This steady state condition persists as long as no sudden changes take place in either the connected supply or the load of the circuit.

Whenever a change of the normal conditions takes place in the electric system, there is a resultant temporary unbalance, and due to the inherent inertia of the system there is a required finite period of time needed by the system to re-establish its previously balanced or steady state condition.

If a fault, in the form of a short circuit current, occurs in an electric system, and if as a result of such fault it becomes necessary to operate an interrupting device; the occurrence of both events, the fault and the current interruption, constitute destabilizing changes to the system that result in periods of transient behavior for the associated currents and voltages.

Interruption of the current in a circuit generally takes place during a transient condition that has been brought about by the occurrence of a short circuit. The interruption itself produces an additional transient that is superimposed upon the instantaneous conditions of the system, and thus it can be recognized that interrupting devices must cope with transients in the currents that have been generated elsewhere in the system, plus voltage transients that have been produced by the interrupting device itself.

2.1 Characteristics of the Short Circuit Current

Because of their magnitude, and of the severity of its effects short circuit currents, undoubtedly, represent the most important type of current transients that can exist in any electrical system.

The main factors that determine the magnitude, and other important characteristics of a short circuit current are, among others, the energy capacity of the source of current, the impedance of the power source, the characteristics of the portion of the circuit that is located between the source and the point of the fault, and the characteristics of the rotating machines that are connected to the system at the time of the short circuit.

The machines, whether a synchronous generator or either an induction or a synchronous motor, are all sources of power during a short circuit condition.

29

Since at the time of the short circuit, motors will act as generators feeding energy into the short circuit due to the inherent inertia of their moving parts.

The combination of these factors and the instantaneous current conditions, prevailing at the time of initiation of the fault will determine the asymmetry of the fault current as well as the duration of the transient condition for this current. These two characteristics are quite important, as it will be seen later, for the application of an interrupting device.

2.1.1 Transient Direct Current Component

Short circuit current transients produced by a direct current source are less complex than those produced by an alternating current source. The transients occurring in the dc circuit, while either energizing, or discharging a magnetic field, or a capacitor through a resistor, are fully defined by a simple exponential function.

In ac circuits, the most common short circuit current transient is equal to the algebraic sum of a transient direct current component, which, as stated before, can be expressed in terms of a simple exponential function, and of a steady state alternating current component that is equal to the final steady state value of the alternating current, and which can be described by a trigonometric function.

The alternating current component is created by the external ac source that sustains the short circuit current. The dc component, in the other hand, does not need an external source and is produced by the electromagnetic energy stored in the circuit inductance.

A typical short circuit current wave form showing the above mentioned components is illustrated in figure 2.1. In this figure the direct current component is shown as I_{DC} , the final steady state component is shown as I_{AC} and the resulting transient asymmetrical current as I_{Total} .

It should be noted that, as a result and in order to satisfy the initial conditions required for the solution of the differential equation that defines the current in an inductive circuit, the value of the direct current component is always equal and opposite to the instantaneous value of the alternating current at the moment of the fault initiation. Furthermore it should be noted that this dc transient current is responsible, and determines the degree of asymmetry of the fault current.

A short circuit current will be considered to be symmetrical when the peak values of each half cycle are equal to each other when measured in reference to its normal axis. An asymmetrical current will be one which is displaced in either direction from its normal axis and one in which the peak value will be different for each half cycle with respect to the normal axis.

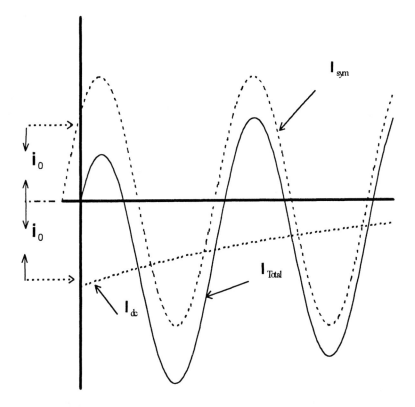

Figure 2.1 Transients components and response of an ac short circuit current.

The total current is mathematically described by the following function:

$$I_{Total} = I_m[(e^{-\alpha t}) \times \sin(\phi) + \sin(\omega t + \phi)]$$

where :

I $_m$ = Peak value of the steady state ac current
α = System Time Constant = R/L
ϕ = fault's initiation angle
R = System Resistance
L = System Inductance

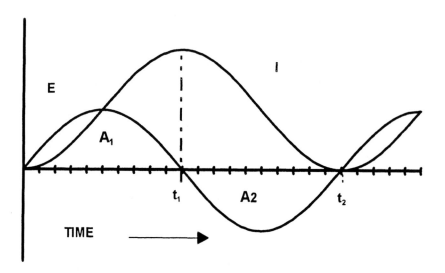

Figure 2.2 Volt-Time and current relationships for a short circuit initiated at the instant of voltage zero.

2.1.2 The Volt-Time Area Concept

To better understand the physics phenomena involving the circuit response to a change of current flow conditions, it is extremely helpful to remember that in a purely inductive circuit the current will always be proportional to the Volt-Time area impressed upon the inductance of the affected circuit. This statement implies that the current is not instantaneously proportional to the applied voltage but that it is dependent upon the history of the voltage application.

The visualization of this concept, in terms of what we are calling the Volt-Time area, is important because it helps to present the inter dependent variations of the voltage and the current at its simplest level. This concept is demonstrated by simply observing the dimensional relationship that exists in the basic formula which describes the voltage appearing across an inductance:

$$e = L \; di/dt$$

Dimensionally this expression can be written as follows:

$$\text{Volts} = \frac{\text{Inductance} \times \text{Current}}{\text{Time}}$$

Solving (dimensionally) for the current we have:

$$\text{Current} = \frac{\text{Volts} \times \text{Time}}{\text{Inductance}}$$

To illustrate the concept, let us consider some specific examples of a fault. In these examples it will be assumed that the circuit is purely inductive, and that there is no current flow prior to the moment of the insertion of the fault.

First, let us consider the case illustrated in figure 2.2 when the fault is initiated at the precise instant when the voltage is zero. In the figure it can be seen that the current is totally shifted above the axis and that the current peak is twice the magnitude of the steady state current. The reason for this is because, in the purely inductive circuit, there is a 90° phase difference between the current and the voltage. Therefore, the instantaneous current should have been at its peak when the voltage is zero; furthermore, as it was stated earlier, the value of the direct current component is equal and opposite to the alternating current at the same instant of time.

It can also be seen that at time t_1 the current reaches its peak at the same moment as the voltage reaches a zero value, this instant corresponds with the point where the area under the voltage curve reverses.

At time t_2 it is noted that both, the current and the voltage reach an instantaneous value of zero. This instant corresponds with the time when the area A_2 under the voltage curve begins its phase reversal, the two areas being equal it then follows that the net area value would be zero, which in turn is also the current value.

The second example, shown in figure 2.3 illustrates the condition when the short circuit current is initiated at the instant where the voltage is at its peak value. At time t_2 it is observed that the area under the voltage curve become negative and therefore the current is seen to reverse until at time t_3 where the current becomes zero since the areas A_1 and A_2 are equal, as it is seen in the corresponding figure.

The first example describes the worst condition, where the short circuit current is fully displaced from its axis and where the maximum magnitude is attained.

The second example represents the opposite, the most benignant of the short circuit conditions, the current is symmetrical about its axis since there is no direct current contribution and the magnitude of the short circuit is the lowest obtainable for all other faults where the voltage and the circuit impedance are the same.

The last example is show in figure 2.4 and is given as an illustration of a short circuit that is initiated somewhere between the voltage zero and the voltage maximum.

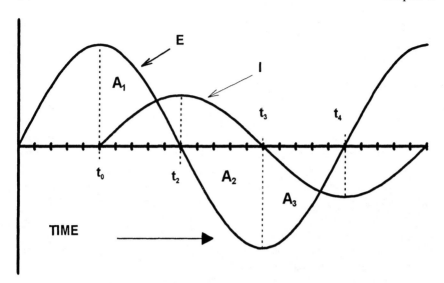

Figure 2.3 Volt-Time and current relationships for a short circuit current initiated at maximum voltage.

One significant characteristic that takes place, and that is shown in the figure is the existence of a major and a minor loop of current about the axis. In this figure we can observe the proportionality of the current and the Volt-Time curve.

It should be noted that the major current loop is the result of the summation of areas A_1 and A_2 and the minor loop corresponds to the summation of A_3 and A_4.

From the above examples and by examining the corresponding figures, the following facts become evident:

1) Peak current always occur at a voltage zero.
2) Current zero always occur when the net Volt-Time areas are zero.
3) The magnitude of the current is always proportional to the Volt-Time area
4) At all points on the current wave the slope of the current is proportional to the voltage at that time.

In all of the examples given above it was assumed that the circuit was purely inductive. This was done only to simplify the explanation of how the short circuit current wave is formed.

In real applications, however, this condition is not always attainable since all reactors have an inherent resistance, and therefore the dc component of the

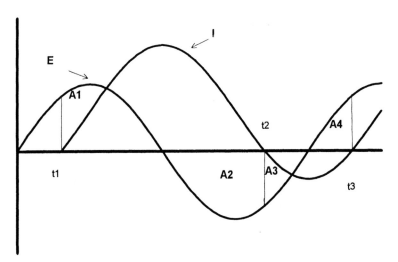

Figure 2.4 Volt-Time and current relationships illustrating major and minor loops of short circuit current.

current will decay exponentially as a result of the electro-magnetic energy being dissipated through this resistance. This condition was illustrated earlier in figure 2.1 where the general form of the short circuit current was introduced.

2.1.3 Transient Alternating Current Components

In the preceding discussions the source of current was assumed to be far removed from the location of the short circuit and therefore the ac current was simply defined by a sinusoidal function. However, under certain conditions the transient ac current may have an additional ac transient component which is the result of changes produced by the short circuit current in the inductance of the circuit.

This condition generally happens when a rotating machine is one of the circuit components that is involved with the short circuit, more specifically it refers to the condition where the contacts making the connection to a generator are closed on a short circuit at the time when the generated voltage passes through its peak. Whenever this happens, the short circuit current rises very rapidly, its rise being limited only by the leakage reactance of the generator stator, or by its sub-transient reactance. This current creates a magnetic field that tends to cancel the flux at the air gap, but to oppose these changes an emf is induced in the generator winding and eddy currents are induced in the pole faces. The net result is that the ac component is not constant in relation to

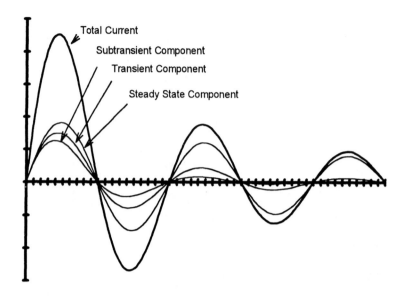

Figure 2.5 The composition of a short circuit current including the ac transient components.

time, but instead it decreases from an initial high value to a constant or steady state value. The particular form of the rate of decrease precludes the use of a single exponential function, and instead it makes it necessary to divide the curve into segments and to use a different exponential expression to define each segment. This results in the use of very distinctive concepts of reactance for each exponential function. The "subtransient" reactance is associated with the first, very rapid decrease period; the "transient" reactance, with the second, less rapid decrease period, and the "synchronous" reactance which is associated with the steady state condition, after all the transients have subsided. In figure 2.5 the transient components as well as the resulting total short circuit current are shown.

2.1.4 Asymmetry of Three Phase Short Circuit Currents

In all of the material that has been presented so far, on the subject of short circuit currents, only single phase system currents were described. The reason for this is because in a simultaneous fault, of a three phase balanced system, or for that matter in any balanced multi-phase system, only one maximum dc component can exist, since only one phase can satisfy the conditions for such maximum at the instant where the fault occurs.

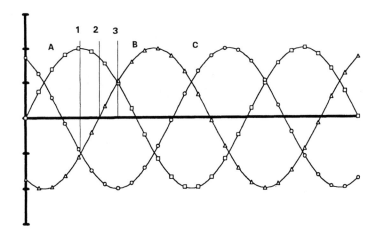

Figure 2.6 Three phase short circuit current characteristics.

Furthermore; a symmetrical short circuit current can not occur in all three phases of a three phase generator, simply because of the current displacement inherent to the a multi-phase system. If a symmetrical fault occurs in one phase, the other two phases will have equal an opposite direct current components, since in all cases the algebraic sum of all of the dc components must be equal to zero.

The preceding statement is demonstrated by referring to figure 2.6, where the three phase steady state currents are shown, and where it can be readily seen that if the peak value of the alternating current is assumed to be equal to 1 in all the phases, and as it has been established before, the initial value of the dc component in a series LR circuit is equal and opposite to the instantaneous value of the ac current which would exist immediately after switching if the steady state could be obtained instantaneously. Observing figure 2.6 we can see that when the initial value of the dc component on phase A is I_m it means that the instantaneous value of the ac current is at its peak, and the steady state value for the currents in phases B and C at the same instant is one half that of the peak value of A.

For a fault that starts at instant 2, there is no offset on phase B since current B is zero at that instant, and we get an offset on phase A that is equal to $-\sqrt{3/2}$, while the offset on phase C is equal to $+\sqrt{3/2}$.

When the short circuit is initiated at time 3, we get a condition that is similar to that which was obtained when the starting point for the fault was at instant 1, except that the maximum offset is observed now on phase C instead of on phase A.

Having shown the conditions where the maximum dc component is produced, let us next look at what happens to the currents some time later. Any combination of dc components that produces a maximum average offset at t = 0 will also produce the maximum offset at any instant thereafter, compared to any other possible combinations of dc components. This follows from the fact that the decay factor of the dc component is identical for the three currents, since it is determined by the physical components of the system which has been assumed to be balanced under steady state conditions. If the dc current decays at the same rate in all three phases then all of the dc components will have decayed in he same proportion from whatever initial value they had. Therefore it follows that the maximum asymmetry of a fault that started with the maximum offset on one phase will occur one half cycle later in the same phase that had the maximum offset.

When considering a fault that has the maximum offset on phase A, we see that the first peak occurs in phase C approximately 60 (or $t = \pi/3\omega$) after the inception of the short circuit. The second peak occurs on phase B at approximately 120 (or $t = 2\pi/3\omega$). The third occurs on phase A at approximately 180 (or at $t = -\pi/\omega$). Since the ac components are identical we can readily tell which peak is larger by comparing the dc components. Starting with phases B and C, we can see that they start at the same value ($I_m/2$). We can also see, without looking at the absolute quantitative values, that since the decay at 60 is less than at 120, the peak of the current on phase C is larger.

Next comparing the peaks of phases A and C, we can see that the dc component on A starts with twice the value of the component of C, but by the time when A reaches its peak it has decayed more than what the C component has.

These relative values can be compared by establishing their ratio.

$$\frac{A}{B} = \frac{I_m e^{-(\pi/\omega T)}}{\frac{I_m}{2} e^{-(\pi/3\omega T)}} = 2e^{-(\pi/\omega T)+(\pi/3\omega T)} = 2e^{-(2\pi R/3X)}$$

From this result we can conclude that the dc components, and therefore the current peaks, would be equal if X/R has a value such that:

$$2e^{-(2\pi R/3X)} = 1$$

When this expression is solved for X/R, which is normally the way in which the time constant of a power system is expressed, we obtain a value equal to 3.02. Which means that for values of $X/R > 3.02$ the value of the exponential is

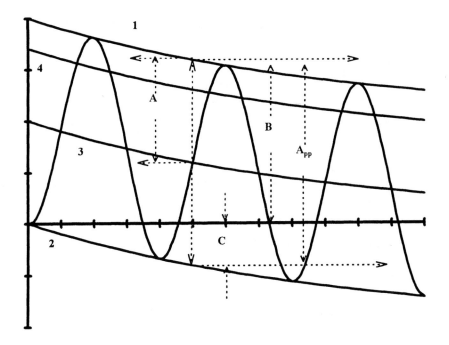

Figure 2.7 Typical asymmetric short circuit current as it can be observed from an oscillogram. The figure illustrates the parameters used for calculating the instantaneous values of current.

greater than 1 and therefore the peak on phase A at 180° will be greater than the peak on phase C at 60°. Conversely for values of X/R smaller than 3.03 the peak on phase A will be smaller than the earlier peak on phase C.

Practical systems, in general can be expected to have X/R ratios significantly greater than 3, and this will cause the peak on phase A, at the end of a half cycle to be the greatest of the three peaks. Being this the most severe peak then it is the one of most interest, as it will be seen later, for circuit breaker applications.

2.1.5 Measuring Asymmetrical Currents

The effective, or rms value of a wave form is defined as the square root of the arithmetic mean of the square of the ordinates of a given curve between two zero points.

Mathematically, for an instantaneous current, which is a function of time, the effective value may be expressed as follows:

$$I_{eff} = \sqrt{\frac{1}{t_2 - t_1} \times \int_{t_1}^{t_2} i^2 dt}$$

In figure 2.7 a typical asymmetric short circuit current during its transient period, as it may be seen in a oscillographic record, is shown. In this figure, during the transient period, the axis of the steady state sinusoidal wave has been offset by an amount equal to the dc component of the short circuit current, and it has been plotted as curve 3. Lines 1 and 2 define the envelope of the total current, and line 4 represent the rms value of the asymmetrical current. The dimension A_{pp} represents the peak to peak value of the sinusoidal wave, A is the maximum value of the current referred to its own axis of symmetry, or in other words the maximum of the ac component (I_M); B and C represents the peak value of the major and minor loop of current respectively.

The general equation for the current wave shown in figure 2.7, which represents a short circuit current where, for the sake of simplicity, the ac transient decrement has been omitted can be written in its simplest form as:

$$i = I_M \sin\phi + D$$

where the term D represents the dc component as previously defined in terms of an exponential function.

As given before, the basic definition of an rms function is given as:

$$I_{rms}^2 = \frac{1}{2\pi} \int_0^{2\pi} i^2 d\theta$$

Now, to obtain an expression that will represent the rms value of the total current we can substitute the first equation, which defines the value of the current as a function of time into the second expression which as indicated represents the definition of the rms value. After simplifying and some manipulation of the trigonometric functions we can arrive to the following expression:

$$I_{eff} = \sqrt{(\frac{I_M}{2})^2 + D^2}$$

In this equation the term I_{eff} is used to represent the rms value of the total current. The term effective is synonymous with rms, however in the context of this derivation, and only for the purpose of clarity this term is used to differentiate between the rms value of the ac component of the current and the rms value of the total current which contains the ac component plus the dc component.

By recognizing that the first term under the radical is the rms value of the ac component itself, and that D represents the dc component of the wave, it can be realized then that the effective, or total rms asymmetrical current is equal to the square root of the sum of the squares of the rms value of the alternating current component and of the direct current component.

If the dc term is expressed as a percent of the dc component with respect to the ac component it becomes equal to $\% \, dc * I_M /100$ and when substituted into the equation for I_{eff} the following expression results.

$$I_{eff} = \sqrt{(\frac{I_M}{2})^2 + (\frac{\%dc}{100})^2 \times I_M{}^2}$$

Furthermore, if in this equation the peak value of the ac component is substituted by the rms value of the same ac component then the equation can be simplified and it becomes:

$$I_{eff} = I_{rms}\sqrt{1 + 2(\frac{\%dc}{100})^2}$$

and this is the form of the equation which is presently shown in the two most influential and significant world wide circuit breaker standards [1] [2].

When a graphical depiction of the wave form of the transient current is available, the instantaneous values, at a time t, for the dc component, and for the rms of the total current can be calculated as follows.

Again referring to figure 2.7 the rms value for the ac component is equal to:

$$I_{rms} = \frac{A}{\sqrt{2}}$$

Furthermore it is seen that:

$$A_{pp} = B + C = 2A$$

Then it follows that:

$$A = \frac{A_{pp}}{2}$$

And in terms of the rms value of the ac component it becomes:

$$\frac{A}{\sqrt{2}} = \frac{A_{pp}}{2\sqrt{2}} = \frac{A_{pp}}{2.828}$$

Also from the figure we find that:

$$A = C + D = \frac{B + C}{2}$$

And solving for the dc component D we get:

$$D = \frac{B - C}{2}$$

And finally the rms value for the asymmetric current can be written in terms of quantities that are directly measurable from the records as:

$$I_{rms} = \sqrt{\frac{A_{pp}}{2.828} + D^2}$$

2.2 Calculation of Short Circuit Currents

Short circuit currents are usually determined by calculations from data about the sources of power and the impedances of all interconnected lines and equipment up to the point of the fault. The precise determination of short circuit currents specially in large interconnected systems, generally requires complicated and laborious calculations. However in most cases there is no need for those complicated calculations and within a reasonable degree of accuracy the approximated magnitude of the short circuit currents can be easily determined within acceptable limits for the application of circuit breakers and the settings of protective relays. Since the resistance in a typical electric system is generally low in comparison to the reactance, it is safe to ignore the resistance and to use only the reactance for the calculations. Furthermore for the calculation of the short circuit current in a system, all generators, and both, synchronous and induction motors are considered as sources of power; load currents are neglected, and when several sources of current are in parallel, it is assumed that all the generated voltages are in phase and that they are equal in magnitude at the time of the short circuit.

While computations of fault currents may be made with reactances expressed in ohms, a number of rules must be observed to take care of machine ratings and of changes in the system voltages that are due to transformers. The calculation will usually be much easier if the reactances are expressed in terms of per cent or per unit reactances. Another simple method of calculation is the MVA method.

2.2.1 The per Unit Method

The per unit method, basically, consists of using the ratings of the equipment as the units for measuring the quantities appearing in problems that involve the same equipment. It is equivalent to adopting a set of units that are tailored to the system under consideration. If we, for example, consider the load on a transformer expressed in amperes, it will not tell us how much we are loading the transformer in relation to what could be considered sound practice. Before we are justified in saying that the load is too much, or is too little, we have to compare it with the normal or rated load for that transformer. Suppose that the transformer in our example is a three phase, 2300-460 volts, 500 KVA transformer and that it is delivering 200 amperes to the load on the low voltage side. The 200 ampere value standing alone does not tell the full story, but when compared to the full load current (627 amperes in this case) it does have some significance. In order to compare these two values we divide one by the other obtaining a value equal to 0.318. This result has more significance than the plain statement of the amperes value of the actual load, because it is a relative measure. It tells us that the current delivered by the transformer is 0.318 times the normal current. We shall call it 0.318 per unit, or 31.8 per cent.

Extending this concept to the other parameters of the circuit, namely; volts, currents, KVAs, and reactances, we can develop the whole theory for the per unit method.

Let V = actual or rated volts
and V_b = base or normal volts
Then, the per unit volts expressed as V_{pu} is defined by:

$$V_{pu} = \frac{V}{V_b}$$

Using a similar notation, we define per unit KVA as:

$$(KVA)_{pu} = \frac{(KVA)}{(KVA)_b}$$

Per unit amperes:

$$I_{pu} = \frac{I}{I_b}$$

and per unit reactance:

$$X_{pu} = \frac{X}{X_b}$$

The normal values for all of the above quantities can be defined with only one basic restriction: The normal values of voltage, current, and KVA must satisfy the following relationship:

$$1000(KVA)_b = V_b \times I_b$$

which when solved for I_b it gives:

$$I_b = \frac{1000(KVA)_b}{V_b}$$

The base or normal values for reactance, voltage, and current are related by Ohm's law as follows:

$$X_b = \frac{V_b}{I_b}$$

and when substituting the value of I_b we obtain:

$$X_b = \frac{V_b^2}{1000(KVA)_b}$$

Which when solving for the per unit value it becomes:

$$X_{pu} = \frac{1000 \times (KVA)_b}{V^2}$$

From the above it follows that a device is said to have a certain per cent reactance when the reactance drop of the device, operating at its rated KVA is that certain per cent of the rated voltage. To describe a reactance, let us say, as 5%, is to say that at rated KVA, or when full load rated current is flowing, the reactive voltage drop is equal to 5% of the rated voltage. Expressing the reactance in per cent is to say that for a rated load, the voltage drop due to the reactance is that number (of volts) per hundred volts of rated voltage. When the reactance is expressed in per unit (pu) this number represents the reactive voltage drop at rated current load per unit of rated voltage.

The per cent and the per unit values when referred to the same base KVA are related by the following simple expression:

$$X\% = X_{pu} * 100$$

When the reactances are given in ohms they can be converted to per unit using the following relationship:

$$X_{pu} = \frac{1000 \times (KVA)_b}{V^2} \times X_\Omega$$

In order that the per cent or per unit values may be used for the calculations of a given circuit, it is necessary that all such values be referred to the same KVA base. The choice for this base KVA can be absolutely arbitrary, it does not need to be tied down to anything in the system. However it will be advantageous to choose as the KVA base a particular piece of equipment in which we are interested.

When the given per cent, or per unit values represent the ratings of a piece of equipment that have a different base than the one that we have chosen as the base KVA for our calculations it will be necessary to translate this information to the same basis. This is readily accomplished by obtaining the proper ratios between the values in question, as it is shown below.

$$\frac{X_{pu}(Base\,2)}{X_{pu}(Base\,1)} = \frac{(KVA)_{b2}}{(KVA)_{b1}}$$

2.2.1.1 General Rules for Use of per Unit Values

The following rules, tabulated below, are given to provide a quick source of reference for calculations that involve the per unit method.

When all reactances are expressed in per unit to the same base KVA, the total equivalent reactance is

a) for reactances in series:

$$X_{total} = X_1 + X_2 \ldots + X_i$$

b) for reactances in parallel:

$$X_{Total} = \cfrac{1}{\cfrac{1}{X_1} + \cfrac{1}{X_2} + \ldots\ldots + \cfrac{1}{X_i}}$$

c) To convert reactances from delta to wye or vice versa:

Delta to Wye

$$X_A = \frac{(X_1)(X_2)}{X_1 + X_2 + X_3}$$

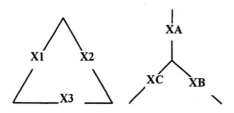

$$X_B = \frac{(X_2)(X_3)}{X_1 + X_2 + X_3}$$

$$X_C = \frac{(X_3)(X_1)}{X_1 + X_2 + X_3}$$

Wye to Delta

$$X_1 = \frac{(X_A X_B) + (X_B X_C) + (X_C X_A)}{X_A}$$

$$X_2 = \frac{(X_A X_B) + (X_B X_C) + (X_C X_A)}{X_B}$$

$$X_3 = \frac{(X_A X_B) + (X_B X_C) + (X_C X_A)}{X_C}$$

2.2.1.2 Procedure for Calculating Short Circuit Currents Using the per Unit Method

The complete procedure used to calculate a short circuit current using the per unit method can be summarized in the following six basic steps:

1. Choose a base KVA
2. Express all reactances in per unit values referred to the chosen base KVA
3. Simplify the circuit by appropriately combining all of the involved reactances. The objective is to reduce the circuit to a single reactance.
4. Calculate the normal, or rated current at the rated voltage, at the point of the fault corresponding to the chosen base KVA.

$$I_n = \frac{(KVA)_b \times 1000}{\sqrt{3} \times V}$$

5. Calculate the per unit short circuit current corresponding to the per unit system voltage divided by the per unit total reactance.

$$I_{pu} = \frac{E = 1}{X_{pu}(total)}$$

6. Calculate the fault current magnitude by multiplying the per unit current (I_{pu}) times the rated current (I_n).

The following example is given to illustrate the application of the per unit method to the solution of a short circuit problem. The circuit to be solved is shown as a single line diagram in figure 2.8.

1. Choose the value that has been given by the utility as the base MVA. This value is 800 MVA.
2. Calculation of the reactances for the different portions of the system yields the following values:

For the 69 kV line $X_{base} = \dfrac{69^2 \times 1000}{80,0000} = 5.95$

For the 13.8 kV line $X_{base} = \dfrac{13.8^2 \times 1000}{80,0000} = 0.24$

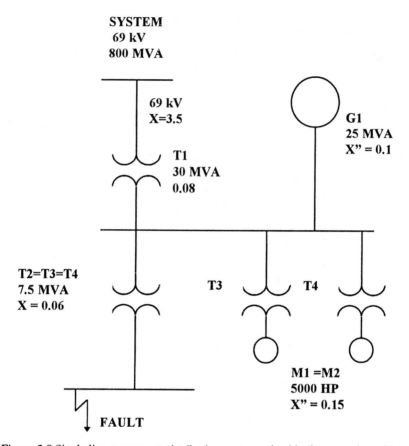

SYSTEM
69 kV
800 MVA

69 kV
X=3.5

T1
30 MVA
0.08

G1
25 MVA
X" = 0.1

T2=T3=T4
7.5 MVA
X = 0.06

T3   T4

M1 =M2
5000 HP
X" = 0.15

FAULT

Figure 2.8 Single line diagram of distribution system solved in the example problem.

For the 4.16 kV line $X_{base} = \dfrac{4.16^2 \times 1000}{800,000} = 0.022$

3. Simplifying the circuit, refer to figure 2.9 (a,b,c,d,e); obtaining the per unit reactances of each component we have as shown in (a):

System (1) = $X1$ = 1.0

Line (2) = $X2 = \dfrac{X}{X_{base}} = \dfrac{3.85}{5.95} = 0.65$

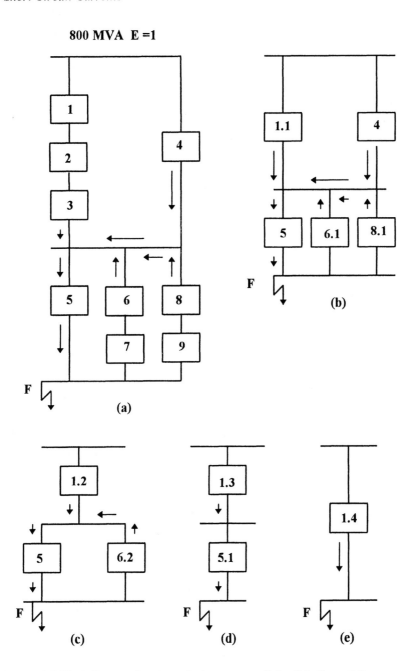

Figure 2.9 Block diagram showing reducing process of circuit in figure 2.8.

Transformer T1(3)

$$= X3 = X_{rated} \times \frac{(KVA)_{base}}{(KVA)_{rated}} = 0.08 \times \frac{800,000}{30,000} = 2.13$$

Generator G1 (4) = $X4 = \dfrac{0.1 \times 800,000}{25,000} = 3.2$

Transformers T2, T3, and T4 (5,6,8)

$$= X5, X6, X8 = \frac{0.06 \times 800,000}{7,500} = 6.4$$

Motors M1, and M2 (7,9) = $X7, X9 = \dfrac{0.15 \times 800,000}{5,000} = 24$

Continuing the process of reducing the circuit, and referring to figure 2.9 (b), the series combination of 1,2,3 is added arithmetically to give:

X1,2,3 (1.1) = 1.0 + 0.65 + 2.13 = 3.78

The two branches containing the components 6, 7 and 8, 9 are also combined independently.

(6.1), (8.1) = 6.4 +24 = 30.4

Next, the just reduced series elements are combined with the parallel components, obtaining the new components which are represented by 1.2 and 6.2 in figure 2.9 (c).

$$(1.2) = \frac{X1,2,3 \times X4}{X1,2,3 + X4} = \frac{3.78 \times 3.2}{3.78 + 3.2} = 1.73 \text{ and,}$$

$$(6.2) = \frac{X6,7 \times X8,9}{X6,7 + X8,9} = \frac{30.4 \times 30.4}{30.4 + 30.4} = 15.2$$

Continuing with the process in figure 2.9 (d) we have:

$$(1.3) = \frac{(1.2) \times (6.2)}{(1.2) + (6.2)} = \frac{1.73 \times 15.2}{1.73 + 15.2} = 1.55$$

and finally we add (1.3) + (5) to obtain the total short circuit reactance:

(1.4) =Xsc =7.95

We then calculate the value of I_{pu} which is equal to:

$$I_{pu} = \frac{E}{X_{sc}} = \frac{1.0}{7.95} = 0.126$$

I_{rated} is calculated as:

$$I_{base(4.16)} = \frac{800,000}{\sqrt{3} \times 4.16} = 111,029$$

Now we can proceed with the calculation of the short circuit current at the specified fault location (F).

$$I_{sc} = I_{pu} \times I_{base} = 0.126 \times 111,029 = 13,966 \text{ Amperes}$$

Alternatively the short circuit current can also be calculated by figuring the equivalent short circuit MVA and then dividing it by the square root times the rated voltage at the point of the short circuit.

$$MVA_{sc} = \frac{MVA_{base}}{X_{pu}} = \frac{800,000}{7.95} = 100.63$$

$$I_{sc} = \frac{MVA_{sc}}{\sqrt{3} \times V_{rated}} = \frac{100.63}{\sqrt{3} \times 4.16} = 13966 \text{ Amperes}$$

2.2.2 The MVA Method

The MVA method is in reality a variation of the per unit method. It generally requires a lesser number of calculations, which makes this method somewhat simpler than the per unit method. The MVA method is based in the fact that when it is assumed that a short circuit current is being supplied from an infinite capacity source, the admittance, which is the reciprocal of the impedance, represents the maximum current, or Volts Amperes at unit voltage, which can flow through a circuit or an individual component during a short circuit condition.

With the MVA method, the procedure is similar to that of the per unit method. The circuit is separated into components, and then these components are reduced until a single component expressed terms of its MVA is obtained. The short circuit MVA for each component is calculated in terms of its own infinite bus capacity. The value for the system MVA is generally specified by the value given by the utility system. For a generator, or a transformer, a line or cable, the MVA is equal to the equipment rated MVA divided by its own impedance, and by the square of the line to line voltage divided by the impedance per phase, respectively. The MVA values of the components are combined according to the following conventions:

1. For components in series:

$$MVA_{Total} = \cfrac{1}{\cfrac{1}{(MVA)1} + \cfrac{1}{(MVA)2} + \ldots \cfrac{1}{(MVA)i}}$$

2. For components in parallel:

$$MVA_{Total} = (MVA)1 + (MVA)2 + \ldots (MVA)i$$

To convert from Delta to Wye or vice versa the same rules that were previously stated in section 2.2.1.1 for the per unit method are applicable.

The same example, shown in figure 2.8, that was solved before, using the per unit method, will now be solved using the MVA method. Referring to figure 2.9 we can use the same schematics while using the per unit method.

The MVA values for each component are calculated as follows:

$$\text{For the system (1) the } MVA = \frac{800,000}{1}$$

$$\text{For the line (2) the } MVA = \frac{69^2}{3.5} = 1,360$$

$$\text{For transformer T1 (3) the } MVA = \frac{30}{0.08} = 375$$

$$\text{For the generator G1 (4) the } MVA = \frac{25}{0.1} = 250$$

$$\text{For the transformers T2,3 and 4 (5,6,8) the } MVA = \frac{7.5}{0.06} = 125$$

For the motors M1,2 (1,2) the $MVA = \dfrac{5}{0.15} = 33.3$

The MVA value for the combined group 1,2,3 is:

$$MVA(1.1) = \frac{1}{\dfrac{1}{800} + \dfrac{1}{1360} + \dfrac{1}{375}} = 465$$

In figure 2.9 (b), the MVA for (6.1) and (8.1) is:

$$MVA\ (6.1),\ (8.1) = \frac{125 \times 33.3}{125 + 33.3} = 26.3$$

In 2.9 (c) the reduced circuit is obtained by combining (6.1) + (8.1) to give ((6.2) whose MVA value is:

$$(6.2)\ MVA = 26.3 + 26.3 = 52.6$$

Next, as shown in (d) the MVA for (1.3) is equal to the parallel combination of (1.2 and (6.2) which numerically is equal to:

$$(1.3)\ MVA = 465 + 52.6 = 518$$

Finally combining the $MVAs$ of (1.3) and (5) we obtain:

$$(1.4) = MVA_{sc} = \frac{518 \times 125}{518 + 125} = 100.69$$

Now the value of the short circuit current can be calculated as follows:

$$I_{sc} = \frac{100.69}{\sqrt{3} \times 4.16} = 13{,}975 \text{ Amperes}$$

It is now left up to the reader to chose whichever method is preferred.

2.3 Unbalanced Faults

The discussion so far has been based under the premise that the short circuit involved all three phases symmetrically, and therefore that this fault had set up a new three phase balanced system where only the magnitude of the currents had changed. However, it is recognized that other than three phase balanced faults can happen in a system; for instance there may be a line to ground or a line to line fault.

Generally it is the balanced three phase fault where the maximum short circuit currents can be observed. In a line to line fault very seldom, if ever the fault currents would be greater than those occurring in the three phase balanced situation. A one line to ground fault obviously is of no importance if the system is ungrounded. Nevertheless in such cases, because of the nature of the fault a new three phase system, where the phase currents, and phase voltages are unbalanced is set up.

Analytical solution of the unbalanced system is feasible but it is usually highly involved and often very difficult. The solution of a balanced three phase circuit, as it has been shown, is relatively simple, because all phases being alike, one can be singled out and studied individually as if it was a single phase. It follows then, that if an unbalanced three phase circuit could somehow be resolved into a number of balanced circuits then each circuit might be evaluated based on its typical single phase behavior. The result of each single phase circuit evaluation could then be interpreted with respect to the original circuit using the principle of superposition. Such tool for the solution of unbalanced faults is afforded by the technique known as the symmetrical components method.

2.3.1 Introduction to Symmetrical Components

The method of symmetrical components is based on Fortesque's [3] theorem, which deals in general with the resolution of unbalanced systems into symmetrical components. By this method an unbalanced three phase circuit may be resolved into three balanced components. Each component is symmetrical in itself and therefore it can be evaluated on the basis of single phase analysis. The three components are: the positive-phase-sequence component, the negative-phase-sequence component, and the zero-phase-sequence component.

The positive-phase-sequence component comprises three currents (or voltages) all of equal magnitude, spaced 120° apart and in a phase sequence which is the same as that of the original circuit. If the original phase sequence is, for example A, B, C, then the positive-phase-sequence component is also A, B, C. Refer to figure 2.10 (a).

The negative-phase-sequence component also comprises three currents (or voltages) all of equal magnitude spaced 120° apart, and in a phase-sequence opposite to the phase sequence of the original vector, that is C, B, A, see figure 2.10 (b).

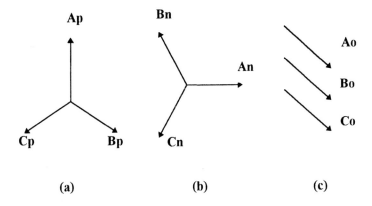

Figure 2.10. Symmetrical vector system, (a) Positive-phase sequence vector; (b) Negative-phase sequence vector; (c) Zero-phase sequence vector.

The zero-phase-sequence component is constituted of three currents (or voltages) all of equal magnitude, but spaced 0° apart, figure 2.10 (c), in this zero-sequence the currents or voltages, as it can be observed, are in phase with each other and in reality they constitute a single phase system.

With the method of symmetrical components, currents and voltages in each phase-sequence interact uniquely; these phase-sequence currents or voltages do not have mutual effects with the currents or voltages of a different phase-sequence, and consequently the systems defined by each phase-sequence may be handled quite independently and their results can then be superimposed to establish the conditions of the circuit as a whole.

Defining impedance as the ratio of the voltage to its respective current makes possible to define impedances that can be identified with each of the phase sequence components. Therefore, there is a positive-phase-sequence impedance, a negative-phase-sequence impedance and a zero-phase-sequence impedance. Each of these impedances may also be resolved into their resistance and reactance components.

The solution of the short circuit using the method of symmetrical components is carried out in much of the same manner as with the per unit or MVA methods. The main difference is that the negative and zero reactances are included in the solution.

In a balanced circuit the negative-phase sequence, and the zero phase-sequence components are absent, and only the positive-phase-sequence is present, therefore the solution is reduced to the simplified methods that were described earlier. In the unbalanced circuit the positive and the negative-phase sequence components are both present, and in some cases the zero-phase-sequence component may also be present. The zero-phase-sequence component will generally be present when there is a neutral or a ground connection. The zero sequence components, however are non-existent in any system of currents or voltages if the vector sum of the original vectors is equal to zero. This also implies that the current of a poly-phase circuit feeding into a delta connection is always zero.

2.3.1.1 Reactances for Computing Fault Currents

As stated earlier, there is a reactance which is identifiable with each phase sequence vector, just as any other related parameter, such as the current, the voltage, the impedance. The reactance can be expressed under the same set of rules described and consequently; the reactances can be identified as follows.

The Positive-Phase-Sequence Reactance, Xp, is the reactance commonly associated and dealt with in all circuits, it is the reactance we are already familiar from our applications of the simplified per unit method. In rotating machinery positive-phase-sequence reactance may have three values; sub-transient, transient and synchronous.

The Negative-Phase-Sequence Reactance, Xn, is present in all unbalanced circuits. In all lines and static devices, such as transformers, the reactances to positive and negative-phase sequence currents is equal, that is, $Xp = Xn$ for such type of apparatus. For synchronous machines and rotating apparatus in general it is reasonable to expect that due to the rotating characteristics, reactances to the positive and the negative-phase-sequence currents will not be equal. In contrast with the possibility of three values for the positive-phase sequence, the negative-phase sequence has but one for rotating machinery, its magnitude is nearly the same as that of the sub-transient reactance for that machine.

The Zero-Phase-Sequence Reactance, Xo, depends not only upon the particular characteristic of the individual device, Which must be ascertain for each device separately. but also it depends upon the way the device is connected. For transformers, for example, the value of the zero-phase-sequence reactance is given not only by the characteristics of its windings but also by the way they are connected.

2.3.1.2 Balanced Three Phase Faults

For balanced three phase faults the value for the total reactance Xt is given by the value of the positive-sequence reactance Xp alone.

2.3.1.3 Unbalanced Three Phase Faults

For unbalanced three phase faults, the same equations may be used with the special understanding as to what the value of Xt is for each case.

Line-to-Line Fault:

$$Xt = \frac{Xp + Xn}{\sqrt{3}}$$

This kind of fault seldom causes a fault current greater than that for a balanced three phase fault because of the usual relationship of Xp and Xn. It is probable, therefore, that investigation of the balanced fault will be sufficient to establish the possible maximum fault current magnitude.

Line-to-Ground Fault.

Obviously this type of fault is of no importance for an ungrounded system, however, if the system is grounded we have the following:

$$Xt = \frac{Xp + Xn + Xo}{3}$$

This equation, usually may be simplified to:

$$Xt = \frac{2Xp + Xo}{3}$$

2.4 Forces Produced by Short Circuit Currents

The electromagnetic forces that are exerted between conductors, whenever there is a flow of current, is one of the most significant, well known, and fundamental phenomena that is produced by the electric current. Bent bus bars, broken support insulators and in many instances totally destroyed switchgear equipment is the catastrophic result of electrodynamic forces that are out of control. Since the electrodynamic forces are proportional to the square of the instantaneous magnitude of the current, then, it is to be expected that the effects of the forces produced by a short circuit current would be rather severe and oftentimes quite destructive.

The mechanical forces acting between the individual conductors, parts of bent conductors, or contact structures within switching devices, can attain magnitudes in excess of several thousand pounds (newtons), per unit length. Consequently the switching station and all of its associated equipment must be designed in either of two general ways to solve the problem.

One way would be to design all the system components so that they are fully capable of withstanding these abnormal forces. The second alternative would be to design the current path in such a way as to make the electromagnetic forces to balance each other. In order to use either approach and to provide the appropriate structures, it is necessary to have at least a basic understanding about the forces that are acting upon the conductors.

The review that follows is intended as a refresher of the fundamental concepts involved. It will also describe a practical method to aid in the calculation of the forces for some of the simplest, and most common cases that are encountered during the design of switchgear equipment. Relatively accurate calculations for any type of conductor's geometric configurations are possible with a piece by piece approach using the methods to be described; however the process will be rather involved and tedious, for more complex arrangements it will preferable to use one of the ever increasing number of computer programs developed to accomplish the task.

2.4.1 Direction of the Forces Between Current Carrying Conductors

This section will be restricted to establish, only in a qualitative form, the direction of the electromagnetic forces in relation to the instantaneous direction of the current. Furthermore, and for the sake of simplicity, only parallel, or perpendicular pairs of conductors will be considered.

To begin with, it would be helpful to restate the following well known elementary concepts:

1. A current carrying conductor that is located within a magnetic field is subjected to a force that tends to move the conductor. If the field intensity is perpendicular to the current, the force is perpendicular to both, the magnetic field, and the current.

The relative directions of the field, the current, and the force are described by Fleming's left-hand rule; which simply says that, in a three directional axis, the index finger points in the direction of the field, the middle finger points in the direction of the current and the thumb point in the direction of the force. In reality this relation is all that is needed to determine the direction of the force on any portion of a conductor where the magnetic field intensity is perpendicular to the conductor.

2. The direction of the magnetic field around a conductor is described by the right-hand rule. This rule requires that the conductor be grasped with the right hand with the thumb extended in the direction of the current flow, the curved fingers around the conductor, then will establish the direction of the magnetic field.

3. Another useful concept to remember is that; the field intensity at a point in space, due to an element of current, is considered as the vector product of the current and the distance to the point. This concept will be expanded in the discussions that are to follow.

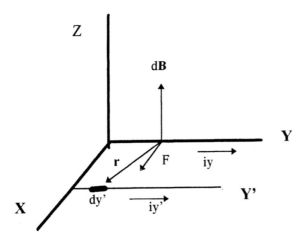

Figure 2.11 Graphical representation of the relationship between current, electromagnetic field, and electromagnetic force for a pair of parallel conductors (Biot-Savart Law).

4. For the purpose of the discussions that are to follow, the directions in space will be represented by a set of coordinates consisting of three mutually perpendicular axis. These axes will be labeled as the X, Y, and Z axis.

2.4.1.1 Parallel Conductors

To begin this review section, the simplest, and most well known case, consisting of two parallel conductors, where each conductor is carrying a current, and where both currents are flowing in the same direction has been chosen. Assuming, as it is shown in figure 2.11, two parallel conductors, Y and Y' carrying the currents i_y and $i_{y'}$. It is also assumed that these currents are both flowing in the same direction, also shown in 2.11. Taking a small portion dy' of the conductor Y', the current in that small portion of the conductor produces a field everywhere in the space around it; therefore, the field intensity produced by that element of current on a point P located somewhere along the second conductor Y can be represented by a vector equal to dB (vector quantities will be represented by bold face characters). Since this vector is the result of the cross product of two other vectors, and furthermore, since the result of a cross vector product is itself a vector, which is perpendicular to each of the multiplied vectors; its direction, when the vectors are rotated in a direction that tends to make the first vector coincide with the second, will follow the direction of the advance of a right handed screw. Applying these concepts it can be seen that the field intensity vector dB at point P must be along a line parallel to the Z axis, since that is the only way that it can be perpendicular to dy' and to **r**, both of which are contained in the X-Y plane. Rotating dy' in a counter clockwise direction will make it coincide with the vector **r**. A right handed screw with its axis parallel to Z would move in the upwards direction when so rotated; then the field intensity d**B** is directed upwards as shown in figure 2.11. The direction that has just been determined for the field intensity can also be verified using the right hand rule.

What has just been determined, for a particular element dy', is applicable to any other element along Y'; therefore, all of the small sections of the conductor Y' produce at point P field intensities that are all acted in the same direction, as previously identified. Furthermore what has been said about a point P on conductor Y, is also applicable to any other point along Y and therefore it can be stated that the field along Y due to the current in Y' is everywhere parallel to the Z axis.

To find the direction of the force that is acting upon the conductor Y, simply apply Fleming's left hand rule, when this is done, it can be seen that the force is parallel to the X axis, as it is shown in the corresponding figure. This result is not surprising and it only confirms a principle which is already well known; that is, two parallel conductors, carrying currents in the same direction attract each other.

If the direction of the current $i_{y'}$ is reversed, the vector product will then yield a field intensity vector d**B** that is directed downwards and then Fleming's rule gives that the force is directed in such a way as to move Y away from Y'. Now, by leaving $i_{y'}$ flowing in its original direction, which will keep d**B** pointing upwards, and reversing the direction of i_y Fleming's rule indicates a force directed so as to move Y away from Y'. Therefore, once again it is verified, the well known, fact that two parallel conductors carrying currents in opposite directions repel each other.

2.4.1.2 Perpendicular Conductors in a Plane

For the arrangement where there are two perpendicular connectors in a plane, it is possible to have four basic cases. For all cases it will be assumed that the conductors coincide with the X and Z axis as shown in figure 2.12.

Case 1. For case number 1, observing figure 2.12 (a), the field d**B** at point P, due to the current i_x in the segment d**x**, is parallel to the Y axis, this is so, because of the result of the vector cross product, since both d**x** and **r** are in the X-Z plane. If d**x** is rotated to make it coincide with the vector **r**, the direction of the rotation is such as to make a right handed screw, parallel to Y to move to the right. This determines the direction of the field, which is readily confirmed by the use of the right hand rule.

The above applies to any d**x** element, and therefore, the total field B at point P coincides with d**B**. Since this applies to any point P on the Y axis, then

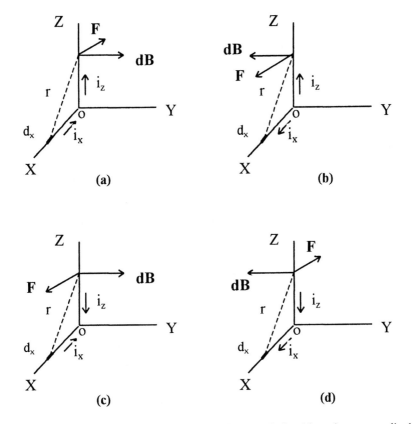

Figure 2.12 Electromagnetic field and force relationships for perpendicular conductors as a function of current direction.

the field at any point on Z, due to the current i_x has the direction shown in figure 2.12 (a). Again, Fleming's left hand rule can be used to determine the direction of the force at any point along the conductor Z , the direction of the force for this particular case is shown in figure 2.12 (a).

Case 2. For case number 2, if the direction of the current i_x is reversed, dB still remains parallel to Y but now it points to the left, as shown on figure 2.12 (b). The force F is then still parallel to the X axis but it is reversed from the direction corresponding to case 1 shown in 2.12 (a).

Case 3. If i_z is reversed, leaving i_x with its original direction, dB is identical to the first case but F is reversed as shown in 2.12 (c).

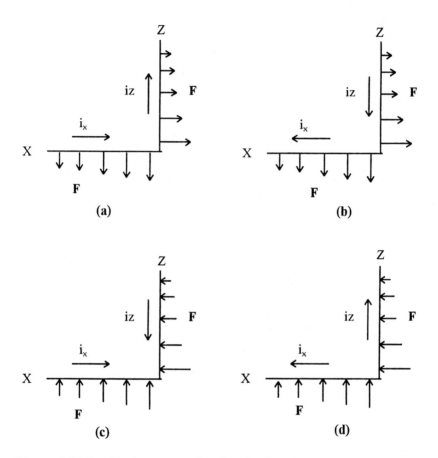

Figure 2.13 Graphical summary showing the direction of the electromagnetic force as a function of current direction

Case 4. Finally if both i_x and i_z are reversed the field d**B** is reversed in comparison to case 1 but the direction of the force remains unchanged as it can be seen on figure 2.12 (d).

The four cases described above have one common feature, that is related to the direction of the current flow. In every case the direction of the force on the Z conductor is such as to try to rotate it about the origin O to make its current, i_z coincide with the current i_x.

The current in Z also produces a force on X. Its direction can be determined by applying the principle just described. A graphical summary

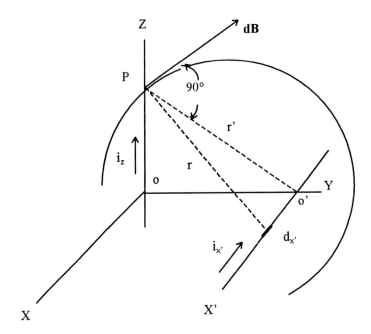

Figure 2.14 Electromagnetic field and forces for a pair of perpendicular conductors not in the same plane.

showing the directions of the forces acting on two conductors at right angles and in the same plane is given in figure 2.13 (a), (b), (c), and (d).

2.4.1.3 Perpendicular Conductors Not in the Same Plane

This particular case is represented in figure 2.14 where it is shown that the field d**B** at point P due to the current i_x is in the Z-Y plane, since it must be perpendicular to d**x** being the vector product of d**x** and **r**. Because of the cross vector product relationship it must also be perpendicular to **r**. Now, a line through point P on the Z-Y plane and perpendicular to **r** is also perpendicular to **r'**, which is the projection of **r** on the Z-Y, (this can be done visually by lifting any of the acute angle corners of a drafting triangle).

The vectors like **r** joining P with all the elements d**x'** along the conductor X' have a common projection **r'** on the Y-Z plane. Therefore the components d**B** at P all coincide, and the total field intensity **B** has a direction as shown in the figure.

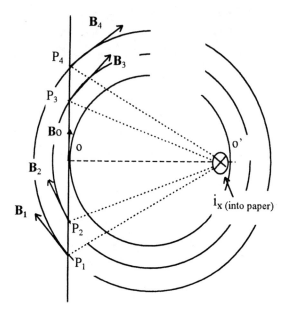

Figure 2.15 Field intensity along the Z axis due to currents in the X' axis.

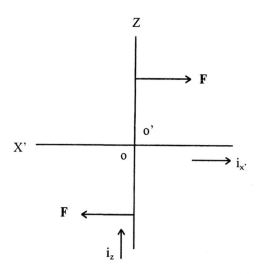

Figure 2.16 Two dimensional view diagram of the force direction. View looking into the end of the Y axis.

The directions of the field intensities at other points along Z, due to the current i_x in X', have directions tangential to circles drawn with their centers at O' as shown on figure 2.15.

The direction for **B** is established by applying the right hand rule. At the point on Z nearest X' the field intensity B_0 coincides in direction with Z. There is no force applied on the conductor at that point, since **B** does not have a component perpendicular to the conductor. At any other point along Z, the field **B** has a component perpendicular to the conductor, and is therefore capable of producing a force on the conductor. In figure 2.15, it can be seen that above O, the nearest point to X' on Z, the perpendicular component of **B** is directed towards X, and below O, away from X.

Putting this information back into a three dimensional diagram the direction of the forces can be determined by applying the left hand rule. Figure 2.16 which is drawn looking into the end of the Y axis, will serve to clarify the results.

The forces on the conductor Z are directed so that they tend to rotate the conductor about OO' so as to make i_x coincide in direction with i_z. If any of the currents are reversed this principle still applies.

2.4.1.4 Direction of Forces Summary

The following observations represent a simplified summary of the results obtained in the preceding section that are related to the determination of the force direction between conductors that are carrying current.

1. The preceding discussion covered parallel runs of conductors, as well as conductors that have rectangular corners, and conductors cross-overs, which may be found in common switchgear construction.
2. In the case of parallel arrangements, the conductors simply attract or repel each other.
3. In the case of perpendicular arrangements, the forces on the conductors simply tend to flip over the conductors so as to make their respective currents to coincide in direction.

Once it is known how a first conductor acts on a second, the basic principle of action and reaction of forces, will define how the second conductor acts on the first; it pushes back when being pushed, and it pulls when being pulled.

2.4.2 Calculation of Electrodynamic Forces Between Conductors

As it is known, the mechanical forces acting on conductors are produced by the interaction between the currents and their magnetic fields. The attraction force between two parallel wires, where both are carrying a common current that is flowing in the same direction is used to define the unit of current, the ampere. According with this definition 1 ampere is equal to the current which will produce an attraction force equal to 2×10^{-7} newtons per meter between two wires placed 1 meter apart.

The calculation of the electrodynamic forces, that are acting, on the conductors is based on the law of Biot-Savart. According to this law the force can be calculated by solving the following equation:

$$\frac{F}{l} = \frac{\mu_0 i_1 \times i_2}{2\pi d} = \frac{\left(4\pi \times 10^{-7}\right) \times i_1 \times i_2}{2\pi \times d} = 2 \times 10^{-7} \frac{i_1 \times i_2}{d} \quad \text{newton per meter}$$

where:

F = Force in newtons
l = length in meters
d = distance between wires in meters
i_1 & i_2 = current in amperes
μ_0 = permeability constant = $4\pi \times 10^{-7}$ (weber / amp-m)

From the above equation, and from the previous discussions, describing the direction of the electromagnetic forces, it is seen that the force between two conductors is proportional to the currents, i_1 and I_2, flowing through the conductors, to the permeability constant μ_0 and to another constant that is defined only by the geometrical arrangement of the conductors. As indicated before, calculations for complex geometric configurations should be made with the aid of any of the many computer models that are available.

The total force acting on a conductor can be calculated as the summation of the component forces that are calculated for each of the individual sections or members of the conductor. However these force components do not exist independently by themselves, since in order to have current flow a complete electric circuit is needed. The use of conductor members in pairs is a convenient way of calculation, but every conducting member in the circuit must be taken in combination with each other member in the circuit.

The force calculation for the bus arrangements that follow only represent some of the most typical, and relatively simple cases that are found in the construction of switchgear equipment.

2.4.2.1 Parallel Conductors

The general equation given by the Biot-Savart law is directly applicable for the calculation of the electromagnetic forces acting between parallel conductors, if the conductors are round and they have an infinite length. The equation is applicable if the ratio between the length of the conductor and the distance

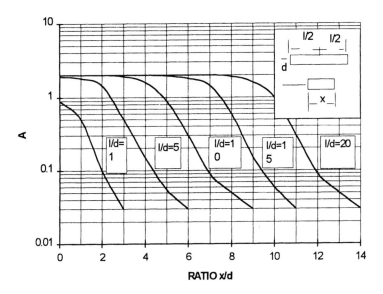

Figure 2.17 Multiplying factor A for calculating the distributed force along a pair of short conductors in parallel.

separating the conductors is more than 10, in which case the resulting error in the calculated force is less than 10 per cent, and therefore it is generally acceptable to use this approximate value for making estimates for the conductor strength requirements.

For conductors of a finite length, the following relationship given by C.W. Frick [4] should be used.

$$F = 2 \times 10^{-7} \times \frac{i_1 \times i_2}{d} \times A \times l$$

where:

$$A = \frac{1}{2l}\left[\sqrt{4d^2 + (l+2x)^2} - \sqrt{4d^2 + (l-2x)^2}\right]$$

The numerical values for the factor A can be obtained from figure 2.17. In the caption box enclosed in figure 2.17 the relationships between the lengths, l and x, of the conductors, and their spacing d, is shown.

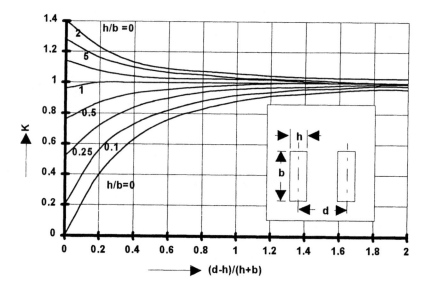

Figure 2.18 Correction factor K for flat parallel conductors.

In the case of rectangular conductors the force can be determined using the same formula that is used for round conductors, except that a shape correction factor K is added to the original formula. This correction factor takes into account the width, thickness, and spacing of the conductors, and accounts for the known fact, that the electromagnetic forces between conductors are not always the same as those calculated under the assumption that the current is concentrated at the center of the conductor. The values for K have been calculated by H.B. Dwight [5] and are given in figure 2.18. For arrangements where the ratio *(d-h) / (h+b) > 2* (refer to figure 2.18) the error introduced is not sufficiently significant and the correction factor may be omitted.

2.4.2.2 Conductors at Right Angles

In figure 2.19 a multiplying factor A is plotted for different lengths l of one of the conductors. The distributed force, per unit length, for the second conductor at a distance y from the bend is calculated using the following formula:

$$\frac{F}{l} = 1 \times 10^{-7} \times i_1 \times i_2 \times A \quad \text{newtons}$$

where:

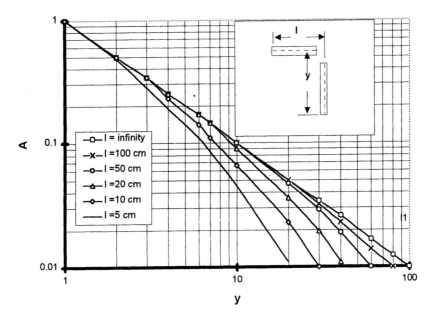

Figure 2.19 Multiplying factor A used to calculate the distributed force along different lengths of conductors at right angles to each other.

$$A = \text{as given in figure } 2.19 = \frac{l}{\left(y\sqrt{l^2+y^2}\right)} \quad \text{and}$$

y and l are as shown in figure 2.19.

The direction of the forces were previously determined in section 2.4 of this chapter. But now for convenience, the general rule will be restated here, as follows:

When in a simple bend the current flows from one leg into the other leg of the bend, the force will act in a direction away from the bend. If the currents flow into the bend from both sides or flow away from the bend from both sides, the forces are directed towards the inside the bend, as if trying to make a straight line out of both conductors.

When one conductor joins another at right angles the force distribution in each conductor can be calculated using the method given above. To find the total effect of the electrodynamic forces, the force on each leg of the conductor is calculated and then the summation of these forces will yield the desired result.

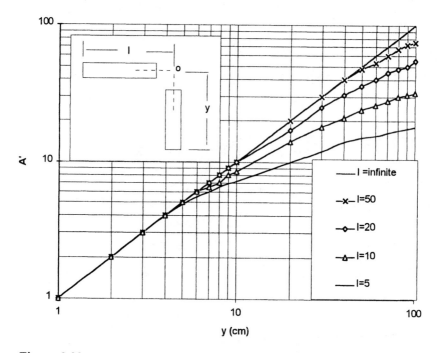

Figure 2.20 Values of A' for calculating the total force at a point on a conductor at right angle.

To calculate the stresses on the individual parts it will be necessary to calculate the force acting on a certain section of the conductor, and in most cases it will also be required to find the moment produced by this force with respect to some given point. The total force acting on the conductor can be calculated using the same equation given for the distributed forces; however, the multiplying factor A will now become A'.

This new factor A' is then defined as:

$$A' = \ln\frac{y}{y_0} \times \left(\frac{l + \sqrt{l^2 + y_0^2}}{l + \sqrt{l^2 + y^2}}\right)$$

where:

l = length of one of the conductor's leg
y = length of the other leg

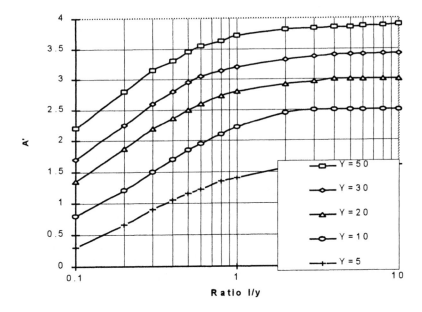

Figure 2.21 Multiplying factor D used for calculating the moment of the force acting on a conductor with respect to an origin point "*o*."

$y_0 = 0.779 \times$ conductor radius (for round conductors) and
$0.224 \times$ (a +b) for square bars where a and b are the cross section dimensions of the bar

Typical values for the multiplying factor A' that have been calculated for different lengths of conductors are shown in figure 2.20.

To calculate the moment of the total force with respect to the center of moments *o*, which in this case corresponds to the center point of the bend, the following equation is used:

$$M_o = 1 \times 10^{-7} \times i_1 \times i_2 \times D$$

where:

D = A multiplying factor whose value is given in figure 2.21

When the origin point of the desired moment is not at the center of the bend, then the moment with respect to a different point can be calculated by first calculating the moment M_o with respect to the point *o*, which can be regarded as the product of the force F and a certain perpendicular distance p, where p is related to o, then the moment of the same force F, with respect to a point at a distance $p+n$ from o will be

$$M_p = F(p+n)$$

where n is the distance from o to p, and therefore

$$M_p = M_o + F \times n$$

Distance n is to be considered either positive or negative according to the position of p with respect to the force and to the point o, that is; n is positive when it is to be added to p and it is negative when it is to be subtracted from p.

2.4.3 Forces on Conductors Produced by Three Phase Currents

As it has been shown before, when a fault occurs on a three phase circuit, because of their difference in phase, all three currents can not be equally displaced,. It is also known that at least two of the currents must be displaced in relation to their normal axis, and in many instances all three phases may well be displaced from the normal axis.

Because of the interaction between all conductors, the forces acting on each of the conductors of a set of three parallel conductors, that are located in the same plane and that are equally spaced can be defined as follows:

The force on conductor 1 is equal to the force due to conductor 2 plus the force due to conductor 3, $(F_1 = F_{1,2} + F_{1,3})$.

The force on conductor 2 is equal to the force due to conductor 3 minus the force due to conductor 1, $(F_2 = F_{2,3} + F_{2,1})$.

The force on conductor 3 is equal to the force due to conductor 1 plus the force due to conductor 2, $(F_3 = F_{3,1} + F_{3,2})$.

When these forces are calculated, and their mathematical maxima is found, assuming that the currents exhibit a phase sequence 1, 2, 3,(when looking at the conductors from left to right), and that the current in phase 1 leads the current in phase 2, while the current in phase 3 lags the current in phase 2, the following generalized results are obtained:

The maximum force on the outside conductors is equal to:

$$F_1 \& F_{3(max)} = 12.9 \times 10^{-7} \left(\frac{i_r^2}{d} \right) \quad \text{newtons per meter}$$

The maximum force on the center conductor is:

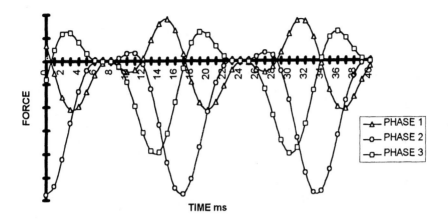

Figure 2.22 Diagram representing the forces on a 3 phase parallel conductor arrangement. Short circuit initiated on phase No.1 at 75 (after current zero in that phase).

$$F_{2(\text{max})} = 13.9 \times 10^{-7} \left(\frac{i_r^2}{d} \right) \quad \text{newtons per meter}$$

where:

i_r = rms value of the symmetrical current

d = center to center distance from the middle to the outside conductors

Figures 2.22 and 2.23 show a plot of the maximum forces for a fault that is initiated in the outside pole, and for one initiated in the center pole, respectively.

It is interesting to note that for any of the three conductors the force reaches its maximum one half of a cycle after the short circuit is initiated. Also it should be noted that the value of the maximum force is the same for either of the outside conductors; however for the force to reach that maximum, the short circuit must occur 75 electrical degrees *before* current zero in the conductor carrying a current that is *leading* the current in the middle conductor, or alternatively it must occur 75 electrical degrees *after* the current zero corresponding to the phase which is *lagging* the current in the middle conductor.

The maximum force on the center phase (conductor 2) will reach a maximum if the short circuit is initiated 45 electrical degrees either, *before* or *after*, the occurrence of a current zero in the middle phase. The maximum

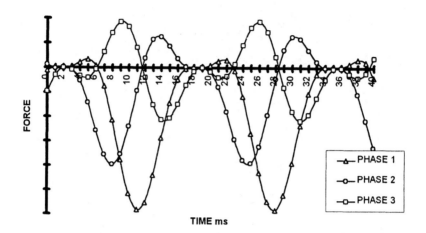

Figure 2.23 Diagram representing the forces on a 3 phase parallel conductor arrangement. Short circuit initiated on center phase (phase No.2) at 45(before current zero in that phase.

force on the middle conductor represents the greatest of the forces on the three conductors.

From all of this, another interesting fact develops; if the short circuit happens 75 degrees *after* the current zero for the current in the conductor that is *lagging* the current in the middle conductor, that instant is 45 degrees *before* the current zero in the middle conductor, therefore under these conditions the respective forces on both of the outside conductors will reach their maximum and this maximum will be reached simultaneously on both phases one half cycle after the initiation of the short circuit. Similarly if the short circuit is initiated 75 degrees *before* current zero in phase 1 that instant corresponds to 45 degrees *after* current zero in phase 2 and therefore the respective forces on phases 1 and 2 will reach their maximum and they will do so simultaneously. one half cycle after the start of the short circuit.

For a conductor arrangement where the spacing is symmetrical, such as in an equilateral triangular arrangement, the maximum value of the resultant force on any conductor will be:

$$F = 13.9 \times 10^{-7} \left(\frac{i_r^2}{d} \right) \quad \text{newtons per meter}$$

The force on any of the conductors in this geometry will reaches its maximum if the instant when the short circuit begins is 90 electrical degrees either *before* or *after* the occurrence of the current zero in that conductor. The force reaches a maximum value under those conditions, 180 electrical degrees, or one half of a cycle after the initiation of the fault. The direction of the maximum force in this case is perpendicular to the plane determined by the other two conductors, and the maximum force is directed away from that plane.

REFERENCES

1. ANSI/IEEE C37.09-1979 Test Procedure for ac High-Voltage Circuit Breakers Rated on a Symmetrical Current Basis.

2. International Standard IEC 56 High Voltage Alternating Current Circuit Breakers. Publication 56: 1987.

3. C. L. Fortesque, Method of symmetrical coordinates applied to the solution of polyphase networks, AIEE Transactions, Vol 37 (Part II): 1027-1140, 1918.

4. C. W. Frick, General Electric Review 36:232-242, May 1933.

5. H. B. Dwight, Calculation of Magnetic Force on Disconnection Switches, AIEE Transactions, Vol 39: 1337, 1920.

6. E. W. Bohnne, The geometry of Arc Interruption, AIEE Transactions Vol 60:524-532, 1941.

7. ANSI/IEEE C37.010-1979 Application Guide for ac High Voltage Circuit Breakers Rated on a Symmetrical Current Basis.

8. J. L. Blackburn, Symmetrical Components for Power Systems Engineering, Marcel Dekker, Inc. 1993.

9. C. F. Wagner, R. D. Evans, Symmetrical Components, McGraw-Hill 1933.

10. M. H. Yuen, Short Circuit ABC-Learn It in an Hour Use It Anywhere, Memorize No Formula, IEEE Transactions on Industrial Applications: 261-172, 1974.

11. B. Bridger Jr., All Amperes Are Not Created Equal: A Comparison of Current Ratings of High-Voltage Breakers Rated According to ANSI and IEC Standards, IEEE Transactions on Industry Applications Vol 29, No. 1: 195-201, Jan.-Feb. 1993.

12. C. N. Hartman, Understanding Asymmetry, IEEE Transactions on Industry Applications Vol 1A-21 No. 4: 842-848, Jul.-Aug. 1985.

13. G. F. Corcoran, R. M. Kerchner, Alternating Current Circuits, John Wiley & Sons. 3rd. Edition, 1955.

14. D. Halliday, R. Resnick, Physics, Part I and II, John Wiley & Sons, 2nd. Ed. 1967.

3

TRANSIENT RECOVERY VOLTAGE

3.0 Introduction

At the beginning of Chapter 2 it was stated that all current interrupting devices must deal with current and voltage transients. Among the current transients, of special interest, are those which are the direct result of sudden changes in the load impedance, such as in the case of a short circuit. Current transients produced by a short circuit are dependent upon events that are part of the system and therefore they are considered as being transients that are induced by the system. The voltage transients, in the other hand, are the result of either the initiation, or the interruption of current flow. These transients are initiated by the switching device itself, and therefore they can be considered as being transients that are equipment induced. However, the characteristics of these transients do not depend on the type of equipment, but rather, they depend upon the parameters, and the specific location of each of the components of the circuit.

What follows is an introduction to this all important subject dealing with voltage transients. Knowledge about the nature and the characteristics of transient voltages, is an essential necessity for all those involved in the design, the application, and the testing of interrupting devices. Transient voltage conditions, especially those occurring following current interruption, must be properly evaluated before selecting an interrupting device, whether it is a circuit breaker, an automatic recloser, a fuse, a load breaking switch, or in general any kind of fault interrupting or load breaking equipment.

Whenever any of the just mentioned devices is applied, it is not sufficient to consider, and to specify only the most common system parameters such as: available fault current, fault impedance ratio (X/R), load current level, system operating voltage, and dielectric withstand levels, but it is imperative that the requirements imposed by the transient voltage be truly understood and properly acknowledged to insure correct application of the selected switching device.

Voltage transients, as stated earlier, generally occur whenever a circuit is being energized, or de-energized. In either case these transients can be quite damaging specially to transformers, reactors and rotating machinery that may be connected to the circuit. The transients occurring during clearing of a faulted circuit and which are referred here as, Transient Recovery Voltages (TRV), will be the first to be considered. In a later chapter other types of voltage transients such as those produced by switching surges, current chopping, restrikes and prestrikes will be covered.

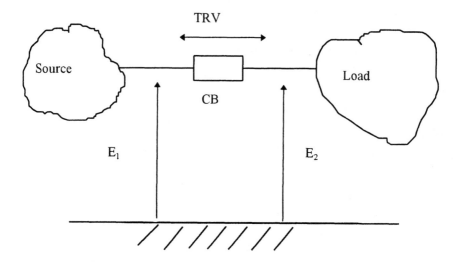

Figure 3.1 Graphical representation of an Electric Network illustrating the sources of the Transient Recovery Voltage.

3.1 Transient Recovery Voltage: General Considerations

All types of circuit interrupting devices can be considered as being a link that is joining two electrical networks. On one side of the device there is the electrical network that is delivering power and which can be identified as the source side network. In the other side there is an electrical network that is consuming power and consequently it can be identified as the load side network, as is illustrated in figure 3.1.

Whenever the interrupting device is opened, the two networks are disconnected and each of the networks proceeds to redistribute its trapped energy. As a result of this energy redistribution, each network will develop a voltage that appears simultaneously at the respective terminals of the interrupter. The algebraic sum of these two voltages then represents the Transient Recovery Voltage, which normally is simply referred as TRV.

A comprehensive evaluation of the recovery voltage phenomena that takes place in any electrical system should be based upon the conditions prevailing at the moment of the interruption of a short circuit current. As minimum requirements to be taken into consideration for this evaluation are: the type of the fault, the characteristics of the network connections, and the switching arrangement used.

Depending upon the different combinations of these conditions, it is obvious that the transient recovery voltage can have many different characteristics; it can exhibit a single frequency, or a multi-frequency response. It can be expressed in the form of a sinusoidal function, a hyperbolic function, a triangular function, an exponential function, or as a combination of these functions; it all depends as it has been said, upon the particular combination of the many factors which directly influence the characteristics of the TRV.

If all factors are taken into consideration exact calculations of the TRV in complex systems is rather complicated, and generally these calculations are best made with the aid of a digital computer program, such as the widely used Electro-Magnetic Transients Program or EMTP.

For those applications where a somewhat less accurate result will suffice, E. Boehne [1], A. Greenwood [2] and P. Hammarlund [3], among others, have shown that it is possible to simplify the calculations by reducing the original system circuits to an equivalent circuit which has a simple mathematical solution. Nevertheless when these simplified calculation method is employed the problem of how to properly select the equivalent circuits and the values of the constants to be used in the calculations still remains.

The selections are only practical equivalents containing lumped components that approximately describe the way in which the actual distributed capacitances and inductances are interrelated in the particular system under consideration. Furthermore the calculation procedures still are somewhat tedious; which again points out the fact that, even for moderately complex systems, it is advantageous to use the modern computer aided methods of calculation.

However, there is something to be said about simple methods for approximated calculations of TRV, and at the risk of oversimplifying the problem, it is possible to say that in the majority of the cases a first hand approximation of the TRV is generally all that is needed for the proper initial selection, and for judging the adequacy of prospective applications of circuit breakers. For some particular cases it is possible to consider just the most basic conditions found in the most common applications. In most cases these are the conditions that have been used as the basis for establishing standards that define the minimum capability requirements of circuit breakers.

A simplified calculation approach can also be of help in determining if the rated TRV of a circuit breaker is sufficient for the application at hand and in many cases the results obtained, with such simplified calculation, can be used to determine if there is a need for further more accurate calculations. Another possible application of the simplified calculation approach is that it can be used to evaluate possible corrective actions that may be taken to match the capability of the device with the characteristics of the circuit. One corrective action is the addition of surge capacitors to modify the inherent TRV of a system.

3.1.1 Basic Assumptions for TRV Calculations

The following assumptions are generally made when calculating the transient recovery voltage of a transmission, or a distribution high voltage power system.

1. Only three phase, symmetrical, ungrounded terminal faults, need to be considered, this is because the most severe TRV appears across the first pole that clears an ungrounded three phase fault occurring at the terminals of the circuit breaker.
2. The fault is assumed to be fed through a transformer, which in turn is being fed by an infinite source. This implies that a fault at the load side terminals of a circuit breaker allows the full rated short circuit current to flow through the circuit breaker.
3. The current flowing in the circuit is a totally reactive symmetrical current; which means that at the instant when the current reaches its zero, the system voltage will be at its peak.
4. The voltage across the circuit breaker contacts, as the current approaches zero, is equal to the arc voltage of the device and it is assumed to be negligible during the TRV calculation, since the arc voltage, when dealing with high voltage circuit breakers, represents only a small fraction of the system voltage. However this may not be the case for low voltage circuit breakers where the arc voltage, in many instances, represents a significant percentage of the system voltage.
5. The recovery voltage rate represent the inherent TRV of the circuit, and it does not include any of the effects that the circuit breaker itself may have upon the recovery voltage.

3.1.2 Current Injection Technique

A convenient contrivance employed for the calculation of the TRV is the introduction of the current injection technique. What this entitles is the assumption that a current equal and opposite to the short circuit current, which would have continued to flow in the event that interruption had not occurred, is flowing at the precise instant of the current zero where the interruption of the short circuit current takes place. Since the currents, at any time, are equal and opposite, it is rather obvious that the resultant value of the sum of these two currents is zero. Consequently the most basic condition required for current interruption is not being violated.

Furthermore it is possible to assume that the recovery voltage exists only as a consequence of this current, which is acting upon the impedance of the system when viewed from the terminals of the circuit breaker.

Additionally, since the frequency of the TRV wave is much higher than that of the power frequency, it is possible to assume, without introducing any

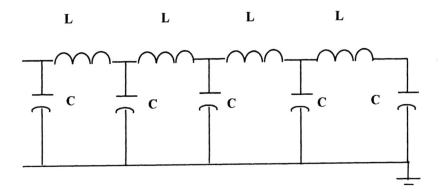

Figure 3.2 Schematic representation of the elements of a Transmission Line.

significant error, that the injected current (i) can be represented by a linear current ramp is defined by:

$$i = \sqrt{2} \times I_{rms} \times \omega \times t$$

where:

I_{rms} = rms value of the short circuit current
$\omega = 2\pi f = 377$
t = time in seconds

As it will be seen later this concept will be used extensively for the calculation of Transient Recovery Voltages.

3.1.3 Traveling Waves and the Lattice Diagram

To better understand some of the important characteristics related to the transient voltage phenomena taking place during the execution of switching operations involving high voltage equipment, it would be beneficial to have at least a basic knowledge about the physical nature and behavior of traveling waves that are present on the transmission lines during these times.

One important characteristic of transmission lines is that since their resistance is generally neglected, they can be represented as being made up of a combination of distributed inductive and capacitive elements. The inductive elements are all connected in series, and the capacitive elements are distributed along the line in parallel as is shown in figure 3.2. When an electrical system is visualized in this fashion, it can be seen that if a voltage is applied to the end of the line, the first capacitor will be charged immediately, and the charging of the capacitors located downstream from the point where the voltage was initially applied will be sequentially delayed as a consequence of the inductors, that are connected in series between the capacitors. The observed delay will be proportionally longer at each point down the line.

If the applied voltage is in the form of a surge signal that starts at zero and that returns to zero in a short time, then it is reasonable to expect that the voltage across the capacitors will reach a maximum value before returning to zero. As this pattern is repeated, at each capacitor junction point along the line, it can easily be visualized that the process serves as a vehicle to propagate the applied surge in the form of a wave which moves along the line. During the propagation of the wave the original characteristics of the surge signal remain basically unchanged in terms of their amplitude and waveform.

Since, in order to charge the capacitors, at each connection point along the line, a current must flow through the inductances that are connecting the capacitors; then, at any point along the line, the instantaneous value of the voltage $e(t)$ will be related to the instantaneous value of the current $i(t)$ by the following relationship:

$$e(t) = Zi(t)$$

Where the constant of proportionality Z represents the surge impedance of the line, which is given by:

$$Z = \sqrt{\frac{L}{C}}$$

In the above relationship L and C are the inductance and the capacitance, per unit length, of the line. The numerical value of the surge impedance Z is a constant in the range of 300 to 500 ohms. A value of 450 ohms is usually assumed for single overhead transmission conductors and 360 ohms for bundled conductors.

Dimensionally, the surge impedance is given in ohms, and is in the nature of a pure resistance; however, it is important to realize that the surge impedance, although it is resistive in nature, it can not dissipate energy as a normal resistive element can. It is also important to note that the surge impedance of a line is independent of the length of such line, this is so because any point located at any distant part of a circuit can not know that a voltage has been applied somewhere in the line until a traveling wave reaches that point.

Traveling waves, as is the case with any other electromagnetic disturbance in air, will propagate at the speed of light, that is 300 meters per microsecond, or approximately 1000 feet per microsecond. As the wave passes from a line that has an impedance equal to Z_1 into another circuit element, possibly but not necessarily another line, which has an impedance equal to Z_2, new waves will propagate from the junction point, traveling back into Z_1, and through the junction into Z_2. The new waves are shaped identically as the incident wave, but their amplitude and possibly their signs are changed.

The coefficients used to obtain the new voltage waves are:

Reflection (from Z_2 back into Z_1):

$$K_R = \frac{(Z_2 - Z_1)}{(Z_2 + Z_1)}$$

Refraction (from Z_1 into Z_2):

$$K_T = \frac{2Z_2}{(Z_2 + Z_1)}$$

If the line termination is a short circuit, then $Z_2 = 0$ and the above equation becomes:

For a reflected wave: $K_{RS} = -1$

For a refracted wave: $K_{TS} = 0$

If the line end is an open circuit, then $Z_2 = \infty$ and the expressions are:

For a reflected wave: $K_{RO} = +1$

For a refracted wave: $K_{TO} = +2$

The back and forward moving waves will pass each other undisturbed along the line, and the potential at any point along such line is obtained by adding the potentials of all the waves passing through the point in either direction.

With the aid of a lattice diagram, figure 3.3, it is possible to keep track of all waves passing through a given point at a given moment. A lattice diagram, can be constructed by drawing a horizontal line from "a" to "b" which repre-

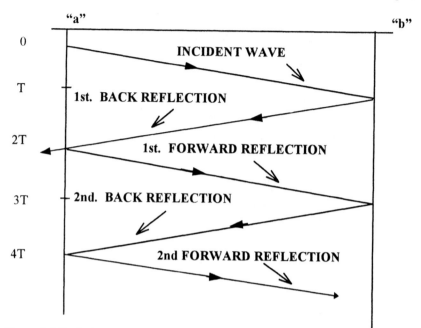

Figure 3.3 Typical construction of a Lattice Diagram.

sents, without any scale, the length of the transmission line. Elapsed time is represented in a vertical coordinate that is drawn downwards from the abscissa, this time is given by the parameter T which symbolizes the time required by the wave to travel from one end of the line to the other end. The progress of the incident wave and of its multiple reflections is then tracked as shown with their corresponding labels by the zigzag lines in figure 3.3.

The next step is to determine the relative amplitudes of the successive reflections. It can be seen that an incident wave, whether it is a current or a voltage wave, which is entering point "a" from the left is a function of time $f(t)$ and it progresses undisturbed to point "b" where the first back reflection takes place. This first back reflection is equal to $K_{Rb}f(t)$, where K_{Rb} is the same coefficient that was earlier defined using the values of Z on both sides of "b". When this wave reaches point "a", the reflection back towards point "b" is obtained by using the coefficient K_{Ra} which also was defined earlier, but, this time the values of Z on both sides of "a" are used. The refraction beyond the point "a" is calculated by the coefficient K_{Ta} which is evaluated using the same relationship given earlier for evaluating the coefficient K_T. The process is repeated for successive reflections, and the amplitude of each successive wave is expressed in terms of these coefficients. The above coefficients can be substituted by the corresponding numerical values defined below and the values can then be used to obtain the actual amplitude of the wave.

For a line terminating on a shorted end: $K_{Rb} = -1$; $K_{Tb} = 0$

For a line terminating on an opened end: $K_{Ra} = +1$; $K_{Ta} = +2$

3.2 Calculation of Transient Recovery Voltages

For any high voltage transmission or distribution network, it is customary and also rather convenient to identify and to group the type of short circuits or faults as either; terminal, or bolted faults and as short line faults.

A terminal fault is one where the short circuit takes place at, or very near, the terminals of the circuit breaker; while a short line fault (SLT) is one where the short circuit occurs at a relatively short distance downstream from the circuit breaker on its load side.

Depending upon the characteristics of the network and the type of the fault, the typical TRV can be represented by either single frequency, or double frequency waveforms for the terminal faults, and by a multi-frequency, that includes a sawtoothed waveform component for the short line fault.

It is also important to recognize that the type of the fault has an important significance on the performance recovery of the circuit breaker. Following a short line fault the circuit breaker is more likely to fail in what is called the thermal recovery region. Which is a region that comprises approximately the first ten microseconds following the interruption of the current, and where thermal equilibrium has not yet been re-established. In figures 3.4 (a) and (b) the oscillograms of a successful and an unsuccessful interruption are included.

For a terminal fault it is more likely that if any failures to interrupt do occur, they will be in the dielectric recovery region, which is the region located between a range from approximately 20 microseconds up to about 1 millisecond depending on the rating of the circuit breaker. In figure 3.5 (a) a dielectric region failure is shown while in figure 3.5 (b) the recovery voltage corresponding to a successful interruption is shown.

3.2.1 Single Frequency Recovery Voltage

A single frequency TRV ensues when during the transient period the electric energy is redistributed among a single capacitive and a single inductive element. In general this condition is met when the short circuit is fed by a transformer and when no additional transmission lines remain connected at the bus following the interruption of the short circuit. This condition generally occurs only in distribution systems at voltages lower than 72.5 kV, where in the ma-

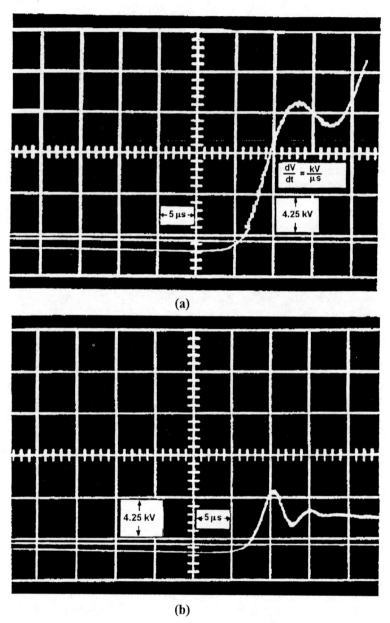

Figure 3.4 Transient Recovery Voltage in the Thermal Recovery Region. Oscillographic traces of a Short Line Fault. (a) Successful interruption. (b) Dielectric failure after approximately 5 μs.

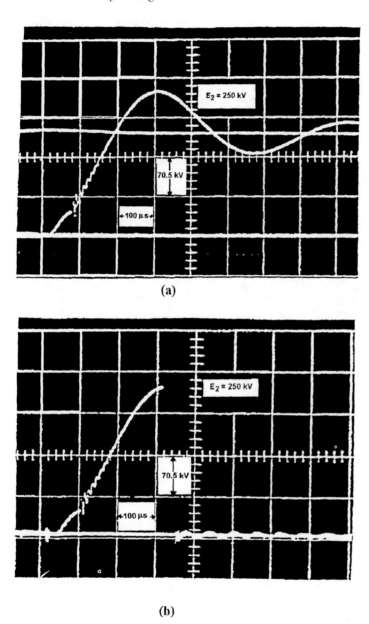

(a)

(b)

Figure 3.5 Transient Recovery Voltage in the Dielectric Recovery Region. Oscillographic traces of a Terminal Fault (a) Successful interruption (b) Dielectric failure after approximately 260 μs.

jority of cases the fault current is supplied by step-down transformers and where because the characteristics of the lines, that are connected to the bus, are such that when considering the transient response of the circuit they are better represented by their capacitance rather than by their surge impedance. As a consequence of this condition the circuit becomes underdamped, and it produces a response which exhibits a typical one minus cosine waveform.

The simplest circuit that serves as an illustration for the single frequency TRV response is shown in figures 3.6 (a) and (b). Referring to this figure, and after opening the switch, the following very basic equation can be written to describe the response of the circuit shown in figure 3.6:

$$V \times \cos\omega t = L\frac{di}{dt} + \frac{1}{C}\int i\,dt$$

where the initial values are:

$I_0 = 0$ since this is a basic requirement for interrupting the current, and

$V_{C0} = 0$ since it was chosen to disregard the value of the arc voltage.

Rewriting the above equation, using the Laplace transform, we obtain:

$$V\left(\frac{S}{S^2 + \omega^2}\right) = SLI_{(s)} + \frac{I_{(s)}}{SC}$$

where S represents the Laplace transform operator.
Solving for the current $I_{(s)}$.

$$I_{(s)} = V\left[\frac{S}{S^2 + \omega^2}\right]\left[\frac{SC}{S^2 LC + 1}\right]$$

Since the TRV is equal to the voltage across the capacitor, this voltage when shown in the Laplace notation is equal to:

$$\mathrm{TRV} = I_{(s)}\left(\frac{1}{SC}\right)$$

Substituting the value of $I_{(S)}$ in the above equation and collecting terms:

$$\mathrm{TRV} = \frac{V}{LC}\left(\frac{S}{S^2 + \omega^2}\right)\left(\frac{1}{S^2 + \dfrac{1}{LC}}\right)$$

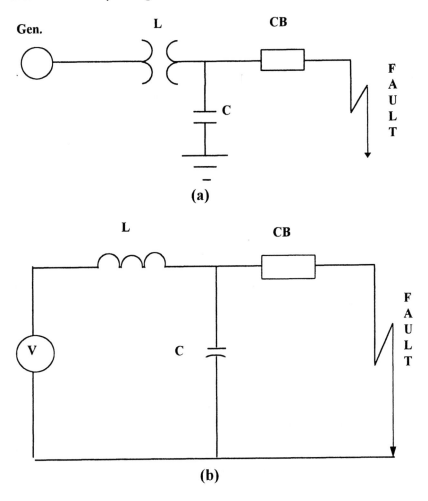

Figure 3.6 Typical simplest circuit which produces a single frequency response.

Letting $\sqrt{\dfrac{1}{LC}} = \omega_0$

and then obtaining the inverse transform, the following equation is obtained:

$$TRV = \frac{V}{LC}\left[\frac{\cos \omega t - \cos \omega_0 t}{\omega_0^2 - \omega^2}\right]$$

And if $\omega_0 \gg$ (then

$$\text{TRV} = V\left(1 - \cos \omega_0 t\right)$$

where:

$$V = E_{TRV} = 1.88 \; E_{rated}$$

The value 1.88 is used as a constant following the recommendations made, for the purpose of standardization, by the Association of Edison Illuminating Companies [4]. This recommendation is based on the fact that at the time of current zero, on an ungrounded three phase terminal fault, the voltage at the source terminal of the breaker is equal to 1.0 per unit while the voltage on the load side terminal is equal to 0.5 per unit so the net steady state voltage across the circuit breaker is equal to 1.5 per unit. However during the transient period, if the effects of any damping are neglected, the voltage can oscillate to a maximum amplitude equal to 3.0 per unit, and therefore it would be reasonable to say that in any practical application the maximum peak of the TRV across the first pole to interrupt a three phase short circuit current could be between 1.5 to 3.0 per unit. When regulation and damping factors, which have been obtained from digital studies, and which have subsequently been verified by field tests were factored in, the final recommended value of 1.88 was chosen.

3.2.2 General Case of Double Frequency Recovery Voltage

In the majority of cases the actual system circuits that are connected to the terminals of a circuit breaker can be represented by relatively simple equivalent circuits composed of lumped capacitive and inductive elements. The substitution of the distributed capacitance and reactance of transformers and generators makes it possible to convert complex circuits into simple oscillatory circuits, which may be easier to handle mathematically. One such system circuit, which is often found in practice is shown in figure 3.7 (a), and its simplified version in 3.7 (b).

As it can be recognized, finding the response of this circuit is not a difficult task since the two frequencies are not coupled together, and in fact, they are totally independent of each other. The solution of this circuit is given by the following relationships that define each one of the two independent frequencies; their resulting waveform which is obtained as the summation of the waveforms that are generated by each independent frequency, and the total voltage amplitude for the recovery voltage.

The frequencies are given by:

$$f_s = \frac{1}{2\pi\sqrt{L_s C_s}} \quad \text{and}$$

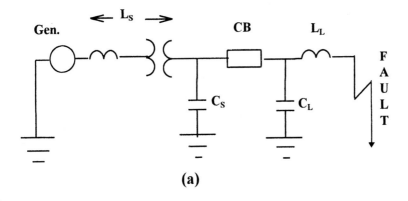

(a)

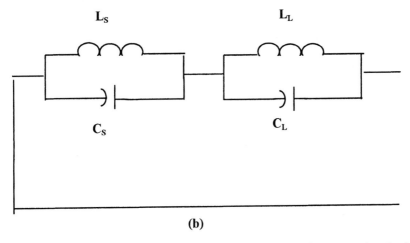

(b)

Figure 3.7 Schematic of simple double frequency circuit used as comparison basis for the calculation of other simplified circuits.

$$f_L = \frac{1}{2\pi\sqrt{L_L C_L}}$$

The magnitude and wave form for the total voltage is proportional to the inductances, and is given by:

$$E_{TRV} = V\left[a_L(1 - \cos\omega_L t) + a_s(1 - \cos\omega_s t)\right]$$

where:

$$a_L = \frac{L_L}{L_L + L_s}; \quad \text{and}$$

$$a_s = \frac{L_s}{L_L + L_s}$$

$$\omega_L = \frac{1}{\sqrt{L_L C_L}};$$

and

$$\omega_s = \frac{1}{\sqrt{L_s C_s}}$$

The above equation, describing the Transient Recovery Voltage of the circuit, is applicable only during the first few hundred microseconds following the interruption of the current, until the power frequency source voltage begins to change by more than a few percent from its peak value. In addition, the equation is accurate only for purely inductive circuits, since this was one of the original assumptions. If the power factor of the power source is such that the phase angle between the current and the voltage is not exactly 90(a more exact expression could be used as shown below.

$$E_{TRV} = V\left[\frac{a_L}{1-\left(\frac{\omega}{\omega_L}\right)^2}(\cos\omega t - \cos\omega_L t) + \frac{a_s}{1-\left(\frac{\omega}{\omega_s}\right)^2}(\cos\omega t - \cos\omega_s t)\right]$$

where:

$\omega = 377$ for a 60 Hz power frequency

3.2.2.1 Circuit Simplification

The aim of the circuit simplification process is to obtain representative circuits from which relationships can be established relating the inductances and the capacitances of these circuits with those of the model circuit described above. For example the relatively complicated network scheme that is shown in figure 3.8 (a) can be simplified as shown in figure 3.8 (b). This simplified circuit then can be mathematically related to the circuit of figure 3.7 by using the following relationships [2],[3].

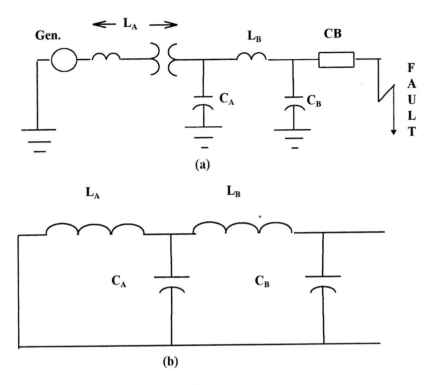

Figure 3.8 Example of circuit simplification.

For the frequency relationships

$$\omega_L = \left(\frac{a-b}{a+b}\right)^{1/4} \left(\omega_A \omega_B\right)^{1/2}$$

$$\omega_s = \left(\frac{a+b}{a-b}\right)^{1/4} \left(\omega_A \omega_B\right)^{1/2}$$

For the amplitude relationships

$$a_L = \frac{\left(a+b-\alpha^2\right)(a+b)}{2(\beta+1)b}$$

$$a_s = 1 - a_L$$

where:

$$a = \frac{1}{2}\left(1 + \beta + \alpha^2\right)$$

$$b = \left(a^2 - \alpha^2\right)^{\frac{1}{2}}$$

$$\alpha = \left(\frac{L_A \ C_A}{L_A \ C_B}\right) = \left(\frac{\omega_B}{\omega_A}\right)$$

$$\beta = \frac{L_A}{L_B}$$

$$\omega_A = \frac{1}{\left(L_A \ C_A\right)^{\frac{1}{2}}}$$

$$\omega_B = \frac{1}{\left(L_B \ C_B\right)^{\frac{1}{2}}}$$

To illustrate the procedure the following numerical example is given by solving the above equations with the following assumed values for the various circuit elements:

L_A = 1.59 mH = source inductance
L_B = 2.39 mH = reactor inductance
C_A = 0.01 μF = capacitance of all equipment on the bus
C_B = 500 pF = capacitance of breaker and reactor

The resultant TRV is shown graphically in figure 3.9

3.2.2.2 Circuit Simplification Procedures

Recognizing that absolute guidelines for simplifying a circuit are not quite possible because one of the major difficulties in the procedure is the proper choice of those circuit components which have a significant influence on the transient phenomena under consideration.

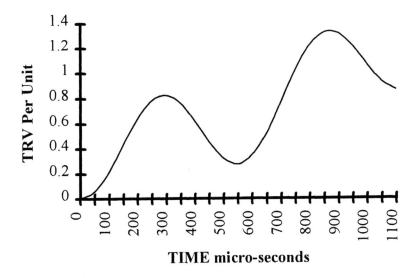

TIME micro-seconds

Figure 3.9 Resultant Transient Recovery Voltage from circuit in figure 3.8.

This selection in many instances is solely based on experience that has been acquired through practice. Despite the limitations, a few generalized rules, that are designed to facilitate the circuit reduction task, can still be provided.

1. The initial circuit is constructed, at least initially, with all the principal components, such as, generators, cables, reactors, transformers and circuit breakers.

2. Starting at the location of the fault and going towards the circuit power source, choose a point where the system is fairly stable. Denote this point as an infinite source, or a generator with zero impedance.

3. The distributed capacitances of generators and transformers are shown as lumped capacitances. The capacitance of each phase of the generator can be substituted by one half of its total capacitance.

4. The inductance of each transformer in the original circuit is replaced by a π type circuit. One half of the total capacitance to ground of the phase winding should be connected at each end of the transformer coil, which corresponds to the leakage inductance of the transformer. The capacitances to ground of both windings must be considered.

5. Whenever two or more capacitances are located in close proximity to each other and are joined by a relatively low impedance, in comparison to the rest of the circuit, the capacitances may be combined into a single equivalent capacitance

6. The value of inductance is calculated from the total parallel reactance of all of the reactances that are connected to the bus bars of the transformers, generators and reactors coils, with the exception of the reactance of the faulted feeder. If cables are used and if they are very short their inductance will be negligible compared with that of the generator and the transformer, and it can be ignored.

7. A capacitance representing the sum of one half of the faulted phase reactor's capacitance to ground, plus one half of the circuit breaker capacitance to ground and the total capacitance to ground of the connecting feeder is connected at the breaker terminals.

8. Show the total capacitance to ground of the cables in the particular phase under consideration. Generally this capacitance is much larger than all others, unless the cables are very short, in which case a capacitance that includes all connected branches and the equivalent capacitances of the reactors, transformers, and generators is considered. This capacitance is always equal to one half of the total capacitance of the circuit components.

9. At the connection point of the generator, the cables and the transformer, the individual capacitances are combined into a single equivalent capacitance.

10. If the transformer is unloaded, the magnetizing inductance of the transformer is much greater than that of the generator, and therefore the generator's inductance can be ignored.

11. If motors are included in the circuit and they are located remote to the fault, their impedances are shown as the single equivalent impedance of all the motors in parallel

3.2.2.3. Three Phase to Ground Fault in a Grounded System

In a three phase system, when the neutral of the system is solidly grounded, and a three phase fault to ground occurs, each phase will oscillate independently it is therefore possible to calculate the response of the circuit using the solution that was previously obtained for a simplified single phase circuit having a configuration as it was shown in figure 3.8 (b).

3.2.2.4 Three Phase Isolated Fault in a Grounded System

As it is already known, the worst case TRV is always observed in the first phase to clear the fault. In the event of an isolated three phase fault occurring within a solidly grounded system, the influence exerted in the recovery voltage by the

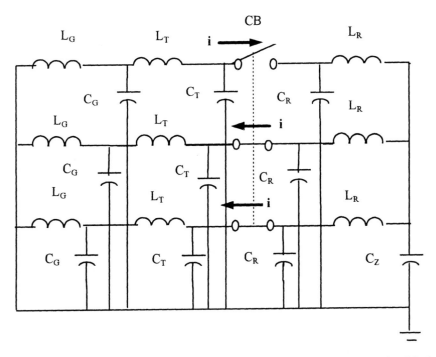

Figure 3.10 Schematic diagram of the equivalent circuit for a three phase isolated fault in a grounded system.

other two phases must be taken into consideration.

In figure 3.10 a representative three phase circuit is shown.. What is significant in this new circuit is the addition of the capacitance shown as C_Z which is equal to the total capacitance to ground from the location of the load inductance to the fault location, this capacitance also includes one half of the capacitance to ground of any reactors present in the system. In figure 3.10 phase A is assumed to be the first phase to clear the fault, and therefore it is shown as being open; while phases B and C are assumed to still be going through the process of interrupting the current and therefore they can be assumed to still be closed. As it can be seen the current flowing through phase A, in the direction of the arrow, will return through the parallel paths of phases B and C, and therefore these two phases can be represented by a single path having one half the inductance and twice the capacitance to ground of the original phases. This is shown in the circuit of figure 3.11 (a).

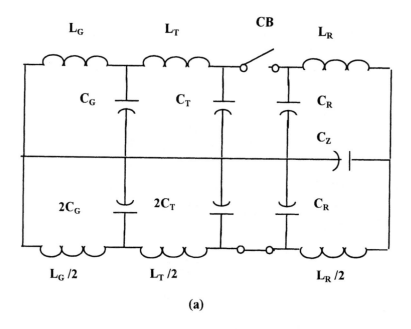

(a)

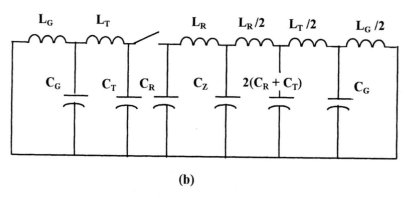

(b)

Figure 3.11 Circuit simplification from original circuit on figure 3.10.

The circuit can be even further simplified as shown in figure 3.11 (b), where it can be observed that the portion of the diagram, to the left hand side of the switch, is composed of two oscillatory circuits which are similar to one of the circuits for which a solution has been provided.

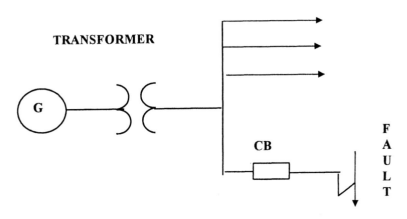

Figure 3.12 Typical system configuration used as basis for defining TRV ratings of high voltage transmission class circuit breakers.

The right hand side of the diagram is made up of four oscillatory circuits, which makes the calculation of this portion of the circuit rather difficult. As a rule, further simplification can generally be achieved by omitting the smallest inductances and, or capacitances and by assuming that such omission will have a negligible effect on the amplitude of the oscillation.

3.2.3 Particular Case of Double Frequency Recovery Voltage

From a more practical point of view of circuit breaker application, it is rather useful to evaluate the transient recovery voltage that appears on a typical transmission circuit which is commonly found in actual high voltage power system applications. This typical circuit is the one that has been used for establishing standard basis of ratings for circuit breakers, the circuit is shown in figure 3.12. As it can be seen in the figure, one of the characteristics of this circuit is that the fault is fed by a source consisting of a parallel combination of one or more transformers and one or more transmission lines.

It is possible to reduce this original circuit to a simpler circuit consisting of a parallel combination of resistive, inductive and capacitive elements.

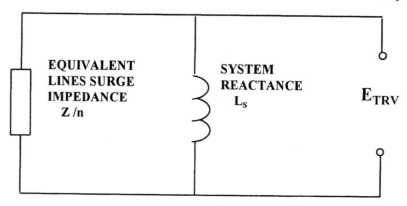

Figure 3.13 Equivalent circuit for the transmission system configuration shown in figure 3.12.

In such a circuit the inductance L is the leakage reactance of the transformer, and the capacitance C corresponds to the total stray capacitance of the installation. The resistance, in this case, represents the total surge impedance Z_n of the transmission lines and is equal to the individual surge impedance Z_o of each line divided by the "n" number of lines interconnected in the system.

In the majority of the applications, the parallel resistance of the surge impedance of the lines is such that it effectively swamps the capacitance of the circuit, and therefore, it is a common practice to neglect the capacitance. The resulting equivalent circuit, which is shown in figure 3.13, is then one that consists of a parallel combination of only inductance (L) and resistance (Z_n).

The operational impedance, for this type of circuit, is given by the following expression:

$$Z_{(s)} = \cfrac{1}{\cfrac{1}{SL} + \cfrac{1}{Z_n}} = \frac{SLZ_n}{Z_n + SL}$$

Now using the injection current technique an expression is obtained for the voltage that appears across the just found circuit impedance. The resulting voltage, which happens to be the TRV of the circuit is given by:

$$V_{(s)} = I_{rms}\sqrt{2} \times \omega \left(\frac{Z_n}{S\left(S + \dfrac{Z_n}{L}\right)} \right)$$

The solution of this equation in the time domain gives the following result:

$$V_{(t)} = \sqrt{2}\left(I_{rms}\right)\omega L\left(1 - \varepsilon^{-\alpha t}\right)$$

where:

$$\alpha = \frac{Z_n}{L}$$

$V_{(t)} = E_{TRV}$ = the exponential component of the total response.

Since the voltage for the first pole to clear is equal to 1.5 times the maximum system voltage then the corresponding transient recovery voltage for this portion of the total envelope is:

$$E_{1(TRV)} = 1.5\sqrt{2}\left(I_{rms}\right)\omega L\left(1 - \varepsilon^{-\alpha t}\right) \text{ or}$$

$$E_{1(TRV)} = 1.5\sqrt{\frac{2}{3}}\ E_{(Rated)}\left(1 - \varepsilon^{-\omega t}\right)$$

where $E_{(Rated)}$ is equal to the rated maximum voltage of the device.

This voltage, which initially appears across the first phase that clears the fault, also appears in the form of a traveling wave; beginning at the bus and traveling down along each of the connected transmission lines. As it is already known, the first reflection of the traveling wave takes place as the result of a discontinuity in the line, from where the traveling wave is reflected back to the breaker terminal, where the traveling wave voltage is added to the initial exponential voltage wave.

Using the lattice diagram method, previously discussed in section 3.1.3, the value of the reflected wave is determined to be equal to the product of the coefficient of reflection K_{Rb}, which is determined by the line termination, times the peak value of the $E_{1\ (TRV)}$ envelope Once again, and according to the traveling wave theory, after the reflected wave arrives to the point where the fault is located it encounters a terminating impedance and its effective value becomes equal to the product of the reflected wave times the coefficient for the refracted wave or $K_{Rb} \times K_{Ta} \times E_{1\ (TRV)}$. When simple reflection coefficients, equal to minus one for a shorted line and plus one for an open line, are used and if the line terminal impedances are resistive in nature; then the coefficients become simply amplitude multipliers.

The TRV calculation can then be reduced to first finding the initial exponential response $E_{1\ (TRV)}$ and then adding, at a time equal to 10.7 microseconds

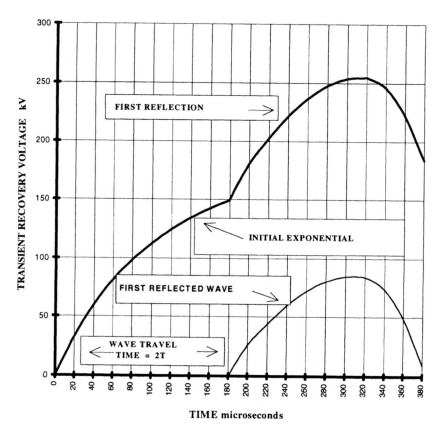

TIME microseconds

Figure 3.14 Typical Transient Recovery Voltage corresponding to the circuit shown in figure 3.13.

per mile, a voltage equal to $K_{Ra} \times K_{Ta} \times E_{1\ (TRV)}$. A typical waveform is illustrated in figure 3.14.

Primarily, to facilitate the testing of circuit breakers that are subjected to this particular form of TRV, The American National Standards Institute (ANSI), has chosen a composite EX-COS model waveform, see figure 3.15, which approximate the actual transient recovery voltage waveform. The initial exponential component of the response is calculated in the same way as before. The equation describing the 1-cosine portion of the response is written as:

$$E_2 = 1.76\ E_{(Rated)}\left(1 - \cos \omega_0 t\right)$$

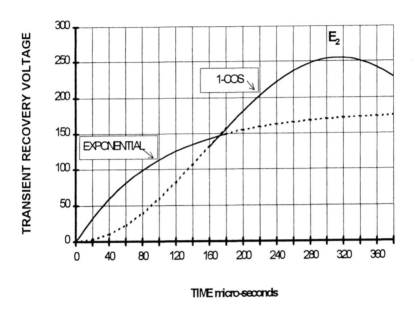

Figure 3.15 Equivalent wave form for testing TRV requirements of circuit breakers.

where:

$$\omega_0 = \frac{\pi}{t_2}$$

The 1.76 multiplier is specified by the American National Standards Institute (ANSI), following the recommendations made by the Association of Edison Illuminating Companies [4]. This multiplier is the product of an statistically collected value of 1.51 times an assumed damping factor of 0.95 and times the now familiar 1.5 factor for the first pole to clear the fault.

The peak value of the voltage is specified by ANSI [5] as the standard requirement for transmission class circuit breakers, (which is the class of circuit breakers rated above 72.5 kV), the standards also specify T_2 as the time in microseconds required to reach the peak of the maximum voltage. Both numerical values are given as a function of the specific rating of the circuit breaker [5].

3.2.3.1 Initial Rate of Rise

The initial slope or initial rate of rise of the TRV, which is also specified in the standards documents is an important parameter that defines the circuit breaker capabilities for it represents one of the limiting values for the recovery voltage. The initial slope is obtained by taking the first derivative of the exponential component of the recovery voltage waveform.

$$\frac{dE_{(TRV)}}{dt} = 1.5 \sqrt{2} \ (I_{rms}\omega L) \left(0 - \varepsilon^{-\alpha t} \left(-\frac{Z_n}{L} \right) \right)$$

Substituting $t = 0$ into the equation the following is obtained,

$$\text{Initial Rate} = R_0 = 1.5 \ \sqrt{2} \ \left(I_{rms}\omega Z_n \right)$$

Inspection of the above equation suggests that the initial TRV rate is directly proportional to the fault current interrupted and inversely proportional to the number of transmission lines that remain connected to the bus. In a sub-transmission class system the feeders are generally in a radial configuration and therefore the fault current does not depend upon the number of connected lines; thus, as the number of lines decreases the fault current remains constant. In the case of transmission class systems, as the number of lines is decreased, so does the fault current.

3.2.4 Short Line Fault Recovery Voltage

A short line fault is a short circuit condition that occurs a short distance away from the load side terminals of a circuit breaker. This short distance is not precisely defined, but it is generally thought to be in the range of several hundred meters up to about a couple kilometers. What makes this type of fault significant is the fact that, as it is generally recognized throughout the industry, it imposes the most severe voltage recovery conditions upon a circuit breaker.

The difficulties arise because the line side recovery voltage appears as a sawtooth wave and therefore the instantaneous, and rather steep initial ramp of voltage imposes severe stresses on the gap of an interrupter before it has had enough time to recover its dielectric withstand capability.

When dealing with the recovery voltage that is due to a short line fault it must be realized that when a fault occurs at some finite distance from the terminals of the protective device, there is always a certain amount of line impedance involved. This line impedance reduces, to some extent, the fault current, but it also serves to sustain some of the system voltage. The further away from the terminals the fault is located, the greater the fraction of the system voltage drop sustained by the line across the load terminals of the circuit breaker.

Consequently, since an unbounded charge can not remain static, then following the interruption of the short circuit current the voltage drop trapped along the line will begin to re-distribute itself in the form of a traveling wave.

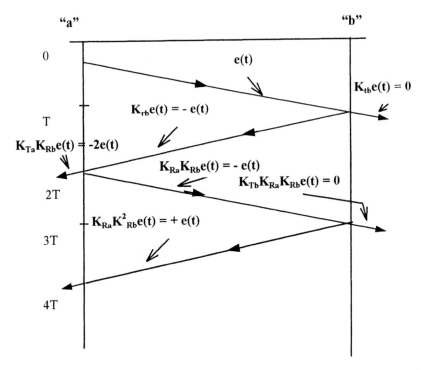

Figure 3.16 Lattice diagram to evaluate characteristics of a short line fault traveling wave.

To evaluate the characteristics of the traveling wave it will be rather convenient to use a lattice diagram, figure 3.16, where the numerical values of the coefficients, for the reflected and transmitted waves are used to shown the amplitude multiplier for the voltage wave as it travels back and forth along the line.

To finally determine the waveform of the line side voltage a simple graphical method, shown in figure 3.17, can be used. In this graph, the horizontal scale represents the time elapsed since the interruption of the current, and is divided in time intervals which are multiples of T; where T is the one way traveling time from the breaker to the location of the fault. On the vertical axis, representing volts at the breaker terminal, the divisions which are set at values corresponding to $E = Z\omega IT$.

When the values from the lattice diagram in figure 3.16 are transferred into figure 3.17, one can notice that the first rising ramp of voltage is a straight line starting at the origin and passing through the locus points $t = 2T$ and $e = 2E$.

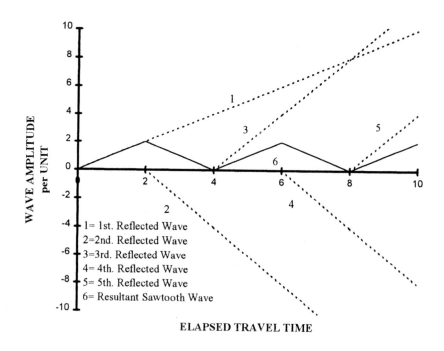

ELAPSED TRAVEL TIME

Figure 3.17 Resultant composite wave for short fault produced by the traveling wave phenomena.

This line has the slope $Z\omega I$ which corresponds to the voltage that is sustained at the breaker terminals until the first reflection returns.

At $t = 2T$ another wave starts from the open end, which according to the lattice diagram of figure 3.16 has a voltage value equal to $-2e_t = -2Z\omega It$. This wave has a zero value at $t = 2T$ and a slope of $-2 Z\omega I$, which is double that of the preceding ramp and in the opposite or negative direction. This ramp is represented by the line drawn from the coordinates $2T$, 0 and $4T$, $4E$.

The lattice diagram further shows that a third wave starts at $t = 4T$ with a positive double slope. Adding up the ordinates of the successive ramps we obtain the expected typical sawtooth wave which characterizes the load side recovery voltage of a short line fault.

The frequency of the response is dependent upon the distance to the fault and to the travel time of the wave. The voltage amplitude is also dependent upon the distance to the fault and on the magnitude of the current.

Specific equations describing the rate of rise and the amplitude of the saw-tooth wave are given by:

$$R_L = \sqrt{2}\, I\, \omega\, Z \times 10^{-6} \text{ kV/}\mu s \text{ and}$$

$$e = dI\sqrt{2}\,(0.58V) \text{ kV}$$

where:

d = an amplitude factor, generally given as 1.6

The time required to reach the voltage peak is given by:

$$T_L = \frac{e}{R_L} \text{ }\mu s$$

The total Transient Recovery Voltage is equal to the sum of the load side transient voltage associated with the traveling wave plus, either the 1 – cosine or the exponential-cosine waveform representing the source side component of the TRV. The choice of the source side waveform depends in the type of system under consideration and is calculated in accordance to the guidelines presented previously.

3.2.5 Initial Transient Recovery Voltage (ITRV)

The term Initial Transient Recovery Voltage (ITRV) refers to the condition where during the first microsecond, or so following a current interruption, the recovery voltage is influenced by the proximity of the system component connections to the circuit breaker. Among those components are the buses, isolators, measuring transformers, capacitors, etc.

Following interruption a voltage oscillation is produced which is similar to that of the short line fault, but, this new oscillation has a lower voltage peak magnitude, but the time to crest has a shorter duration due to the close distance between the circuit breaker and the system components.

The traveling wave will move wave will move down the bus up to the point where the first discontinuity is found. In an IEC report [6], the first discontinuity is identified as being the point where a bus bar branches off, or the point where a capacitor of at least one nanofarad is connected.

The following expression for the ITRV is given in reference [12].

$$E_i = \omega\sqrt{2}\,IZ_b\,T_i\,10^{-6} \text{ }kV$$

where

Z_b = Surge impedance = 260 Ohms
T_i = Wave travel time in microseconds
I = Fault current kA
$\omega = 2\ \pi f$

The above expression shows that:

1. The first peak of the ITRV appears at a time equal to twice the traveling time, of the voltage wave, from the circuit breaker terminals to the first line discontinuity.
2. The initial slope of the ITRV depends only on the surge impedance of the bus and the rate of change of current (di/dt at I=0).

The above statements suggest that since the travel time is a function of the physical location of the component, it is practically impossible to define a general form of ITRV. One can expect that there would be as many ITRV variations as there are station layouts.

Nevertheless, representative values for various voltage installations have been established [7]. These values are tabulated in table 3.1.

TABLE 3.1

Rated Maximum Voltage kV rms	121	145	169	242	362	550	800
Time to First Voltage Peak T_i (s)	0.3	0.4	0.5	0.6	0.8	1.0	1.1

The second item, deals with the rate of change of current. It suggests that if the slope of the current is modified by the action of the breaker during interruption then it can be expected that the ITRV would also be modified.

All this indicates that at least in theory the ITRV exists. However in practice there are those who question their existence saying that the ideal breaker assumed for the calculations does not exist.

It appears that it is better to say that there may be some circuit breakers that may be more sensitive than others to the ITRV. More sensitive breakers are those that characteristically produce a low arc voltage and that have negligible or no post arc current, in other words, an ideal circuit breaker.

REFERENCES

1. E. W. Boehne, The Determination of Circuit Recovery Rates, AIEE Transactions, Vol 54; 530-539. 1935.
2. Allan Greenwood, Electrical Transients in Power Systems, John Wiley and Sons Inc., 1971.
3. P. Hammarlund, Transient Recovery Voltage Subsequent to Short Circuit Interruption, Proc. Royal Swedish Academy Engineering Sci. No. 189, 1946.
4. Transient Recovery Voltage on Power Systems, Association of Edison Illuminating Companies, New York, 1963.
5. ANSI C37.06-1979, Preferred Ratings and Related Required Capabilities for ac High Voltage Circuit Breakers Rated on a Symmetrical Current Basis.
6. IEC 56-1987 High-voltage alternating-current circuit-breakers.
7. ANSI/IEEE C37.04-1979 Standard Rating Structure for AC High-Voltage Circuit Breakers Rated on a Symmetrical Current Basis.
8. ANSI/IEEE C37.011-1979 Application Guide for Transient Recovery Voltage for ac High Voltage Circuit Breakers Rated on a Symmetrical Current Basis.
9. O. Naef, C. P. Zimmerman, J. E. Beehler, Proposed Transient Recovery Voltage Ratings for Power Circuit Breakers, IEEE Transactions, Power Apparatus and Systems, Vol 76 Part III; 1508-1516, 1958 .
10. C. L. Wagner, H. M. Smith, Analysis of Transient Recovery Voltage (TRV) Rating Concepts, IEEE Transactions on Power Apparatus and Systems, Vol PAS 103, No. 11; 3354-3363, April 1984 .
11. R. G. Colclaser Jr., D. E. Buettner, The Traveling Wave Approach to Transient Recovery Voltage, IEEE Transactions on Power Apparatus and Systems, Vol PAS 86; 1028-1035, June 1969.
12. S. R. Lambert, Application of Power Circuit Breakers, IEEE Power Engineering Society, 93 EHO 388-9-PWR; 32-37, 1993.
13. G. Catenacci, Electra 46, 39 (1976).

4

SWITCHING OVERVOLTAGES

4.0 Introduction

Among the most common reasons for dielectric failures in an electric system, aside from lighting strikes, are the overvoltages produced by the switching that is normally required for the ordinary operation of the electrical network.

Switching overvoltages can be produced by closing an unloaded line, by opening an isolating switch, or by interrupting low currents in inductive or capacitive circuits where the possibility of restrikes exists.

Switching overvoltages are probabilistic in nature and their appearances in a system depend mainly upon the number of faults that must be cleared on a line and on how frequently routine switching operations are performed on a particular system. This implies that not only opening operations that are intended for interrupting a short circuit current are responsible for switching overvoltages; but, also the many routine operations that are performed, sometimes daily, in a system. These routine operations are fully capable of producing overvoltage effects by virtue of them altering the system configuration.

As it has been said repeatedly, overvoltages in transmission and distribution systems can not be totally avoided, but their effects can be minimized. Generally the occurrence and the magnitude of the overvoltage can be limited by the use of appropriate measures such as the use of series or parallel compensation, closing resistors, surge suppressors; such as metal oxide varistors, or snubbers containing combinations of resistors and capacitors, and in some cases by simply following basic established procedures for the proper design and operation of a system. [1]

It is appropriate at this point to emphasize that although circuit breakers participate in the process of overvoltage generation they do not generate these voltages, but rather these voltages are generated by the system. Circuit breakers, however can provide means for decreasing, or controlling, these voltage surges. They can do so either by timing controls or by incorporating additional hardware such as closing resistors as an integral part of the circuit breaker design. [2]

4.1 Contacts Closing

The simple closing of a switch or of a circuit breaker can produce significant overvoltages in an electric system. These overvoltages are due to the system adjusting itself to an emerging different configuration of components as a result of the addition of a load impedance. Furthermore, there are charges that are trapped in the lines and in the equipment that is connected to the system and these charges now must be re-distributed within the system.

In addition, and whenever the closure of the circuit occurs immediately after a circuit breaker opening operation the trapped charges left over from the preceding opening can significantly contribute to the increase in the magnitude of the overvoltages that may appear in the system. It is important to note that in most cases the highest overvoltages will be produced by the fast reclosing of a line. It should also be realized that the higher magnitudes of the overvoltages produced by the closing or the reclosing operation of a circuit breaker will always be observed at the open end of the line.

Although the basic expressions describing the voltage distribution across the source and the line are relatively simple, defining the effective impedance that controls the voltage distribution within the elements of the circuit is rather difficult and generally can only be adequately handled with the aid of a computer. [3]

Because of the complexity of the problem no attempt will be made here to provide a quantitative solution. The aim of this chapter will be to describe qualitatively the voltage surges phenomena that take place during a closing or a reclosing operation, and during some special cases of current interruption. The upper limits of overvoltages that have been obtained either experimentally or by calculation will be quoted but only as general guidelines.

4.1.1 Closing of a Line

A cable that is being energized from a transformer represents the simplest case of a switching operation as is shown in figure 4.1 (a). For the sake of simplicity, the transformer has been represented by its leakage inductance; while the cable is represented by its capacitance. As a result of this simplification the equivalent circuit can take the form of the circuit illustrated in figure 4.1 (b).

The transient voltage, shown in figure 4.1 (c), oscillates along the line at a relatively low single frequency and has an amplitude that reaches a peak value approximately equal to twice the value of the system voltage that was present at the instant at which the closure of the circuit took place.

Although the above described circuit may be found in some very basic applications, in actual practice it is more likely to expect that a typical system will consist of one or more long interconnected overhead lines, as depicted in figure 4.2 (a). The equivalent circuit, and the transient response of this system is shown in figure 4.2 (b) and (c) respectively. The transient response, as can be seen in the figure, is determined by the combined impedance of the transformer

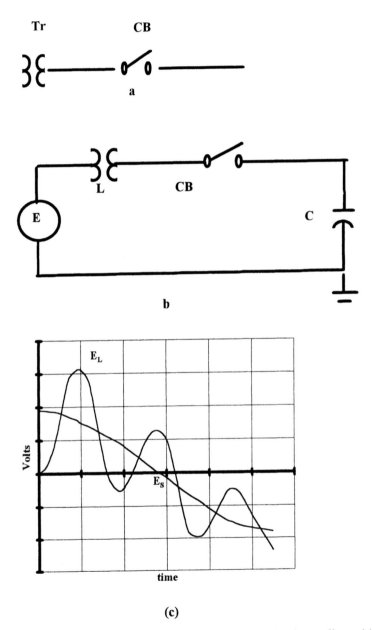

(c)

Fig 4.1 Representation of the simplest case of closing into a line. (a) Single line schematic, (b) Equivalent circuit and (c) Transient surge.

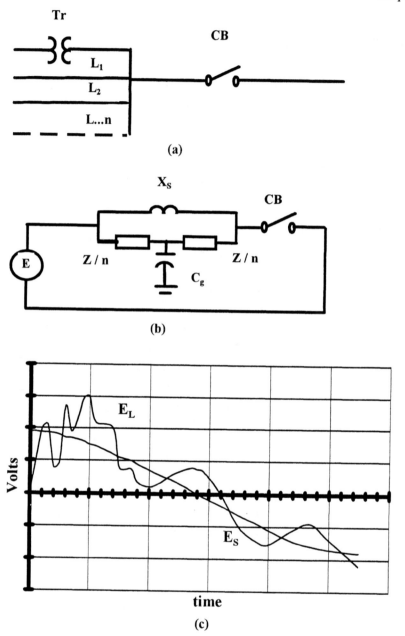

Figure 4.2 Switching surge resulting from energizing a complex system. (a) Single line schematic of the system, (b) Equivalent circuit and (c) Surge voltage.

that is feeding the system and by the total surge impedance of the connected lines. The total surge impedance, as it can be recalled, is equal to the surge impedance of each individual line divided by the number of connected lines.

The total closing overvoltage is given by the sum of the power frequency source overvoltage and the transient overvoltage being generated at the line.

The overvoltage factor for the source is given by the following equation.

$$K_S = \frac{1}{\cos 2\pi f \sqrt{LC}l - \dfrac{X_s}{Z} \sin 2\pi f \sqrt{LC}l}$$

where:

    f = power frequency
    L = positive sequence inductance per length of line
    C = positive sequence capacitance per length of line
    l = line length
    X_s = short circuit reactance of source
    Z = surge impedance of the line

It is evident, by simply observation of the above equation, that a higher power frequency overvoltage factor can be expected as a result of the following occurrences:

1. When the length of the lines increase
2. When the source reactance increases
3. When the surge impedance of the line is lowered as a direct result of an increased number of connected lines and
4. When the power frequency is increased, which means that the overvoltage is higher in a 60 Hz system than in a 50 Hz one.

The overvoltage factor for the transient response portion of the phenomena is not as easy to calculate manually and a simple formula as in the preceding paragraph is just not available. However, it is possible to generalize and it can be said that the overvoltage factor for the transient response is proportional to:

1. the instantaneous voltage difference between the source voltage and the line voltage as the contacts of the circuit breaker close,
2. the damping impedance of the lines connected at the source side of the circuit and
3. the terminal impedance of the unloaded line/ lines being energized

In any case what is important to remember is:

1. When switching a number of lines the amplitude factor of the overvoltage is always reduced as the size of the system increases, and

2. The reduction of the amplitude factor is not due to the damping effects of the system but rather to the superposition of the individual responses each having a different frequency.

4.1.2 Reclosing of a Line

Since in order to improve the stability of the system it is desirable to restore service as quickly as possible, it is a common operating practice to reclose a circuit breaker a few cycles after it has interrupted a fault.

If the interrupted fault happens to be a single phase to ground fault, then it is possible that a significant voltage may remain trapped in the unfaulted phases. This happens because the three phases represent a capacitor that has been switched off at current zero and therefore, because of the inductive nature of the system, this coincides with the instant where a maximum voltage is present in the line.

Since the closing of the contacts may take place at any point in the voltage wave, it could then be expected that when reclosing the circuit, the circuit breaker contacts may close at the opposite polarity of the trapped charge, which, when coupled with the voltage doubling effect produced by the traveling wave, leads to the possibility of an overvoltage across the contacts that can reach a magnitude as high as 4 per unit.

4.2 Contact Opening

The opening of a circuit was previously discussed in the context of interrupting a large magnitude of current where that current was generally considered to be the result of a short circuit. However, there are many occasions where a circuit breaker is required to interrupt currents that are in the range of a few amperes to several hundred amperes, and where the loads are characterized as being either purely capacitive or purely inductive.

The physics of the basic interrupting process; that is the balancing of the arc energy is no different whether the interrupted currents are small or large. However, since lower currents will contribute less energy to the arc it is natural to expect that interrupting these lower currents should be a relatively simpler task; but, this is not always the case because, as it will be shown later, the very fact that the currents are relatively low in comparison to a short circuit current promotes the possibility of restrikes occurring across the contacts during interruption. Those restrikes can be responsible for significant increases in the magnitude of the recovery voltage.

According to standard established practice, a restrike is defined as being an electrical discharge that occurs one quarter of a cycle or more after the initial current interruption. A reignition is defined as a discharge that occurs not later than one eighth of a cycle after current zero.

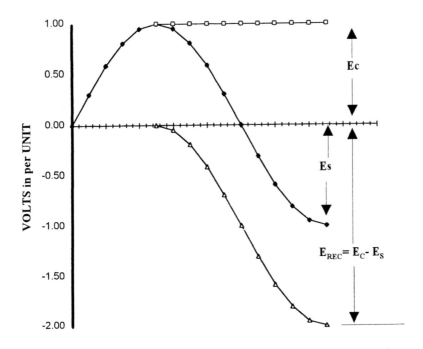

Figure 4.3 Recovery voltage resulting from the switching of capacitor banks

4.2.1 Interruption of Small Capacitive Currents

The switching of capacitor banks and unloaded lines requires that the circuit breaker interrupts small capacitive currents. These currents are generally less than ten amperes for switching unloaded lines and most often less than one thousand amperes for switching off capacitor banks.

Interruption, as always, takes place at current zero and therefore the system voltage, for all practical purposes is at its peak. This as it should be recalled makes current interruption relatively easy but, again as it was said before, this is not necessarily so because those low currents may be interrupted when the gap between the circuit breaker contacts is very short and consequently, a few milliseconds later as the system recovery voltage appears across the circuit breaker contacts the gap is still rather small and it may be very difficult for the circuit breaker to withstand the recovery voltage.

At the time when current interruption takes place the line to ground voltage stored in the capacitor in a solidly grounded circuit is equal to 1.0 per unit.

The source side, in the other hand, will follow the oscillation of the power frequency voltage and therefore in approximately one half of a cycle the voltage across the contacts would reach its peak value but, with a reversal of its polarity. At this time then the total voltage across the contacts reaches a value of 2.0 per unit which corresponds to the algebraic sum of the capacitor voltage charge and the source voltage as is shown in figure 4.3.

If the circuit has an isolated neutral connection then the voltage trapped in the capacitor, for the first phase to clear, has a line to ground value of 1.5 per unit and the total voltage across the contacts one half of a cycle later will then be equal to 2.5 per unit.

Restrikes can be thought as being similar to a closing operation where the capacitor is suddenly reconnected to the source, and therefore it is expected that there will be a flow of an inrush current which due to the inductance of the circuit and in the absence of any damping effects will force the voltage in the capacitor to swing with respect to the instantaneous system voltage to a peak value that is approximately equal to the initial value at which it started but with a reversed polarity. If the restrike happens at the peak of the system voltage, then the capacitor voltage will attain a charge value of 3.0 per unit. Under these conditions, if the high frequency inrush current is interrupted at the zero crossing, which some circuit breakers are capable of doing so, then the capacitor will be left with a charge corresponding to a voltage of 3.0 per unit and one half of a cycle later there will be a voltage of 4.0 per unit applied across the circuit breaker contacts. If the sequence is repeated, the capacitor voltage will reach a 5.0 per unit value, as is illustrated in figure 4.4. Theoretically, and if damping is ignored, the voltage across the capacitor can build up according to a series of 1, 3, 5, 7, . . . and so on without limit.

4.2.2 Interruption of Inductive Load Currents

When a circuit breaker that has an interrupting capability of several tens of kiloamperes is called upon to interrupt inductive load currents that are generally in the range of a few tens to some hundreds of amperes, as for example in the case of arc furnace switching, those currents are interrupted in a normal fashion, that is at current zero. However, and again due to the high interrupting capacity of the circuit breaker those small inductive currents can be interrupted rather easily in a manner similar to that which was described in connection with the interruption of small capacitive currents.

At the time of interruption the gap between the contacts may be very short, and since the voltage is at its peak, then in many cases the small gap may not be sufficient to withstand the full magnitude of the recovery voltage which begins to appear across the contacts immediately following the interruption of the current. As a consequence the arc may restrike resulting in a very steep voltage change and in significant overvoltages.

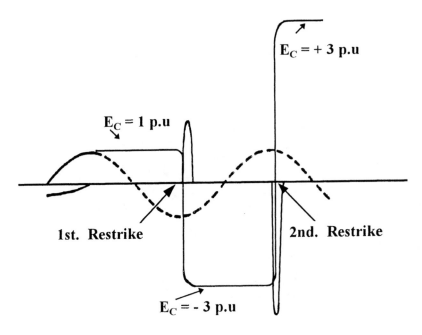

$E_C = + 3$ p.u

$E_C = 1$ p.u

1st. Restrike

2nd. Restrike

$E_C = - 3$ p.u

Figure 4.4 Voltage escalation due to restrikes during a capacitance switching operation.

However, because of the randomness of the point at which the restrikes take place and due to the inherent damping of the circuit, it is very unlike that the upper limit of these overvoltages will exceed a value of 2 per unit.

There are however special cases that arise when a circuit breaker has exceptional capabilities for interrupting high frequency currents, such as those generated by a reignition or a restrike. Whenever the high frequency current is interrupted the normal power frequency recovery voltage reappears across the contacts and in some cases it is possible that a restrike may occur again.

During the interval between the two reignitions the contacts have moved thus increasing the gap distance and therefore a higher breakdown voltage is to be expected. Nevertheless, during this interval more magnetic energy is accumulated in the inductance of the load and consequently additional energy is available to trigger a breakdown which would occur at a voltage that is higher than the previous one. This process may repeat itself as successive reignitions occur across a larger gap and at increased magnetic energy levels, and

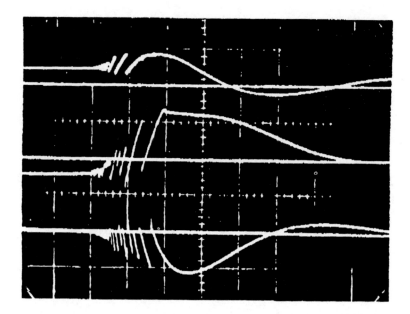

Figure 4.5 Voltage surges caused by successive reignitions when interrupting low inductive currents.

therefore, at higher mean voltage levels resulting in a high frequency series of voltage spikes such as those shown in figure 4.5.

Because of the statistical nature of this phenomenon it is not possible to establish an upper limit for the overvoltage; however, it is advisable to be aware of the potential risk and to use protective devices such as surge arrestors.

4.2.3 Current Chopping

Current chopping is the result of the premature extinction of the power frequency current before a natural current zero is reached that is due to the arc instability as the current approaches zero.

It is commonly believed that only vacuum circuit breakers are capable of chopping currents. However, this is not the case, all types of circuit breakers can chop. Nevertheless, what is different is that the instantaneous current magnitude at which the chopping occurs varies among the different types of interrupting mediums and indeed it is higher for vacuum interrupters. [4], [5]

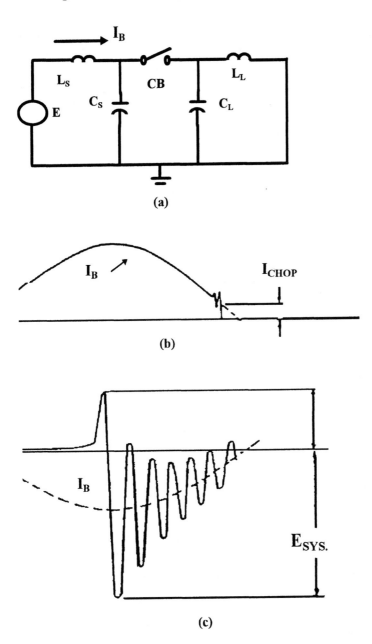

Figure 4.6 Typical Current Chopping. (a) Equivalent circuit, (b) Chopped current across the breaker and (c) Transient voltage across the breaker.

In theory, when current chopping occurs the current is reduced instantaneously from a small finite value to zero, but, in reality this does not happen so suddenly simply because of the inductance that is present in the circuit and as it is well known, current can not change instantaneously in an inductor. It is therefore, to be expected that some small finite element of time must elapse for the transfer of the magnetic energy that is trapped in the system inductance.

At the instant when current chopping occurs the energy stored in the load inductance is transferred to the load side capacitance and thus creating a condition where overvoltages can be generated. In figure 4.6 (a) the simplified equivalent circuit is shown and in (b) the voltage and current relationships are illustrated.

Referring to the equivalent circuit the energy balance equations can be written as:

$$\frac{1}{2}CE_m^2 = \frac{1}{2}CE_c^2 + \frac{1}{2}LI_0^2$$

and the overvoltage factor K is given by:

$$K = \frac{E_m}{E_s} = \sqrt{\left(\frac{I_0}{E_s}\right)^2\left(\frac{L}{C}\right) + \frac{E_c}{E_s}}$$

where:

E_m = Overvoltage peak
E_0 = Peak voltage at supply side
E_c = Capacitor voltage at instant of chop
I_0 = Instantaneous value of chopped current

$\frac{L}{C}$ = Surge impedance of the circuit

As it can be seen, the magnitude of the overvoltage factor K is highly dependent upon the instantaneous value of the chopping current.

4.2.3.1 Current Chopping in Circuit Breakers other than Vacuum

For air, oil, or SF_6 interrupters, the arc instability that leads to current chopping is primarily controlled by the capacitance of the system. The effects of the system capacitance on the chopping level are illustrated in figure 4.7 [6]. The effects of the capacitance on vacuum interrupters is also included in this figure for comparison purposes.

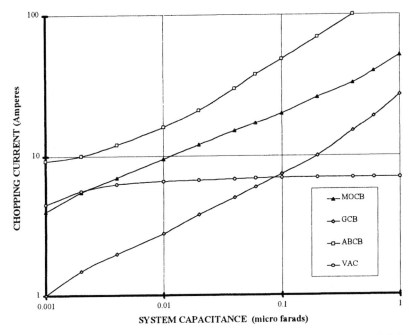

Figure 4.7 Current Chopping Level as Function of System Capacitance for Minimum Oil Circuit Breakers (MOCB), SF$_6$ Gas Circuit Breakers (GCB), Air Blast Circuit Breakers (ABCB), and Vacuum Circuit Breakers (VAC).

For gas or oil circuit breakers the approximate value of the chopping current is given by the formula

$$I_0 = \lambda \sqrt{C_L}$$

where:

λ= Chopping number

The following are typical values for chopping numbers:

For Minimum Oil circuit breakers	7 to 10 × 10^4
For Air Blast circuit breakers	15 to 40 × 10^4
For SF$_6$ Puffer circuit breakers	4 to 17 × 10^4

The values of the system capacitance can be assumed to be in the range of 10 to 50 nano-farads.

4.2.3.2 Current Chopping in Vacuum Circuit Breakers

In contrast to other types of circuit breakers the current instability in vacuum interrupters is not strongly influenced by the capacitance of the system (see figure 4.7), but is dependent upon the material of the vacuum contacts and by the action of the anode spot created by the vacuum arc. There is no chopping number for vacuum interrupters but, instead the chopping current itself can be specified as follows:

> For Copper-Bismuth contacts current chopping   5 to 17 Amperes
> For Chrome Copper contacts current chopping   2 to 5 Amperes

4.2.4 Virtual Current Chopping

Virtual current chopping in reality is not a true chopping phenomenon but rather it is the normal interruption of a fast transient current. Virtual chopping is a phrase that has been coined to describe the condition illustrated in the simplified circuit shown in figure 4.8.

Referring to this figure, the power frequency currents are shown as I_A, I_B and I_C. Assuming that for example, a current reignition occurs shortly after the interruption of the power frequency current in phase A, the reignition current i_a will then flow to ground through the line to ground capacitance C_g in the load side of the breaker in phase A and the components i_b and i_c flow in phases B and C due to the coupling of their respective line to ground capacitances.

The high frequency transient current produced by the reignition superimposes itself on the power frequency; furthermore, the high frequency current could be larger in magnitude than the power frequency current and therefore it can force current zeroes at times other than those expected to occur normally with a 50 or 60 Hz current.

As it has been stated before there are some types of circuit breakers which are capable to interrupt these high frequency currents, and therefore it is possible to assume that in some cases the circuit breaker may clear the circuit at a current zero crossing that has been forced by the high frequency current and that the zero crossing occurs at a time prior to that of the natural zero of the power frequency current. When this happens, as far as the load is concerned, it looks the same as if the power frequency current has been chopped since a sudden current zero has been forced.

Since the high frequency current zeroes will occur at approximately the same time in all three phases the circuit breaker may interrupt the currents in all three phases simultaneously thus giving rise to a very complicated sequence of voltage transients that may even include reignitions in all three phases.

Considering that, when compared in a "normal" current chopping, we find that the instantaneous value of current, from which the load current is forced to

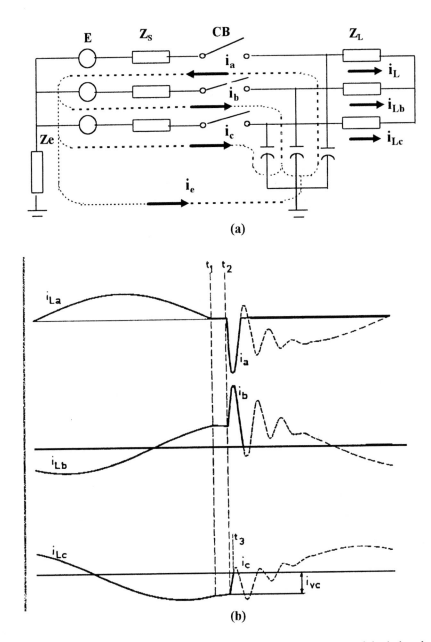

(a)

(b)

Figure 4.8 Virtual current chopping (a) Circuit showing the flow of the induced currents. (b) Relationships between the three phase currents (from Ref.7).

zero, is significantly higher but, also that the surge impedance is somewhat lower, then the line to ground overvoltage could be assumed to be at about the same order of magnitude as the overvoltages that are generated by the conventional current chopping; however, in the worst case, if the neutral is ungrounded one half of the reignition current would return through each of the other two phases and they both will be in phase but with opposite polarities which will result in the line to line overvoltage of the two phases being twice their corresponding line to ground overvoltage.

4.2.5 Controlling Overvoltages

Circuit breakers themselves do not generate overvoltages, but they do initiate them by changing the quiescent conditions of the circuit. As it has been stated before the switching overvoltages are the result of two overvoltage components the power frequency overvoltage, and the transient overvoltage component. Limiting the magnitude of the first is usually sufficient to reduce the total overvoltage to within acceptable limits. However, this does not exclude the possibility of using appropriate measures to additionally limit the magnitude of the overvoltage by limiting the transient response.

Among the measures that can be taken to reduce the magnitudes of the power frequency overvoltages are:

(a) Provide polarity controlled closing
(b) Add closing and or opening resistors across the breaker contacts
(c) Provide a method combining polarity control and closing resistors
(d) Add parallel compensation
(e) Reduce the supply side reactance

The transient overvoltage factor can be controlled by:

(a) removing the trapped charges from the line
(b) synchronized closing which can be accomplished either by closing at a voltage zero of the supply side or by matching the polarity of the line and the supply side.
(c) synchronized opening which optimizes the contact gap at current zero
(d) using pre-insertion resistors

From all the listed alternatives only resistors can be considered to be an integral part of a circuit breaker. The practice of including closing resistors as part of a circuit breaker is relatively common for circuit breakers intended for applications at voltages above 123 kV.

REFERENCES

1. Allan Greenwood, Electrical Transients in Power Systems, Wiley-Interscience, New York, 1971.

2. R. G. Colclaser Jr., Charles L. Wagner, Edward P. Donahue, Multistep Resistor Control of Switching Surges, IEEE Trans. PA&S Vol. PAS-88 No. 7, July 1969, 102-1028.

3. G. W. Stagg, A. H. El-Abiad, Computer methods in Power System Analysis, McGraw Hill New York 1968.

4. S. Berneryd, Interruption of Small Inductive Currents Simple Physical Model and Interaction with Network. The University of Sydney. Abstracts, Symposium on Circuit Breaker Interruption and Power Testing, May 1976.

5. Small inductive Current Switching, CIGRE WG 13.02, Electra 72 .

6. M. Murano, S. Yanabu, H, Ohashi, H. Ishizuka and T. Okazaki, Current Chopping Phenomena of Medium Voltage Circuit Breakers IEEE Trans. PA&S Vol. PAS-97 No.1, Jan-Feb 1977, 143-149.

7. Virtual current chopping, Electra, 72, 87-90.

8. Thomas E. Brown, Circuit Interruption, Theory and Applications, Marcel Dekker Inc., New York, 1984.

9. W. S. Meyer, T. H. Liu, Electromagnetic Transients Program Rule Book, Bonneville Power Administration Portland Oregon, 1980.

5

TYPES OF CIRCUIT BREAKERS

5.0 Introduction

Circuit breakers are switching devices which according to American National Standards Association (ANSI) C37.100 [1] are defined as: "A mechanical device capable of making, carrying and breaking currents under normal circuit conditions and also making, carrying for a specific time and breaking currents under specified abnormal circuit conditions such as those of short circuit."

Historically, as the operating voltages and the short circuit capacities of the power systems have continued to increase, high voltage, high power circuit breakers have evolved trying to keep pace with the growth of the electric power systems. New technologies, primarily those involving the use of advanced interrupting mediums, have been developed and continue to be studied [2].

To achieve current interruption some of the early circuit breaker designs simply relied on stretching the arc across a pair of contacts in air; later arc chute structures, including some with magnetic blow-out coils were incorporated, while other devices used a liquid medium, including water but more generally oil, as the interrupting medium.

Some of those early designs have been significantly improved and variations of those types of circuit breakers are still in use specially in low voltage applications where presently plain air circuit breakers constitute the dominant type of circuit breakers used.

For indoor applications, at system voltages that range from approximately 5 kV up to 38 kV, air magnetic circuit breakers were the breakers of choice in the US until the mid nineteen seventies, while in Europe and other installations outside of the US, minimum oil circuit breakers were quite popular. Bulk oil and air blast circuit breakers where quite common until the mid seventies for outdoor applications at voltages ranging from 15 kV up to 345 kV. For indoor, medium voltage applications minimum oil circuit breakers enjoyed a great deal of popularity in Europe.

With the advent of vacuum and sulfurhexafluoride the older types of circuit breaker designs have been quickly superseded and today they can be effectively considered as being obsolete technologies.

What we just described suggests that circuit breakers can be used for different applications, that they can have different physical design characteristics and that they also can perform their interrupting duties using different quenching mediums and design concepts.

5.1 Circuit Breaker Classifications

Circuit breakers can be arbitrarily grouped using many different criteria such as; the intended voltage application, the location where they are installed, their external design characteristics, and perhaps and most importantly, by the method and the medium used for the interruption of current.

5.1.1 Circuit Breaker Types by Voltage Class

A logical starting point for establishing a classification of circuit breakers is the voltage level at which the circuit breakers are intended to be used. This first broad classification divides the circuit breakers into two groups:

1. Low voltage circuit breakers, which are those that are rated for use at voltages up to 1000 volts, and
2. High voltage circuit breakers. High voltage circuit breakers are those which are applied or that have a rating of 1000 volts or more.

Each of these groups is further subdivided and in the case of the high voltage circuit breakers they are split between circuit breakers that are rated 123 kV and above and those rated 72.5 kV and below. Sometimes these two groups are referred as the transmission class, and the distribution or medium voltage class of circuit breakers, respectively.

The above classification of high voltage circuit breakers is the one that is currently being used by international standards such as ANSI C37.06 [3], and the International Electrotechnical Commission (IEC) 56 [4].

5.1.2 Circuit Breaker Types by Installation

High voltage circuit breakers can be used in either indoor or outdoor installations. Indoor circuit breakers are defined in ANSI C37.100 [1] as those "designed for use only inside buildings or weather resistant enclosures."

For medium voltage breakers ranging from 4.76 kV to 34.5 kV it generally means that indoor circuit breakers are designed for use inside of a metal clad switchgear enclosure.

In practice, the only differences between indoor and outdoor circuit breakers is the external structural packaging or the enclosures that are used, as illustrated in figures 5.1 and 5.2. The internal current carrying parts, the interrupting chambers and the operating mechanisms, in many cases, are the same for both types of circuit breakers provided that they have the same currents and voltage ratings and that they utilize the same interrupting technology.

5.1.3 Circuit Breaker Types by External Design

From the point of view of their physical structural design, outdoor circuit breakers can be identified as either dead tank or the live tank type circuit breakers. These two breaker types are shown in figures 5.3 and 5.4.

Figure 5.1 Indoor type circuit breaker.

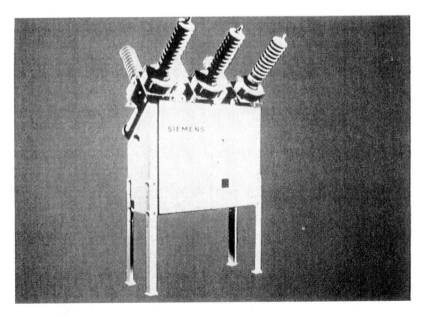

Figure 5.2 Outdoor type circuit breaker.

131

Dead tank circuit breakers are defined in ANSI C37.100 [1] as "A switching device in which a vessel(s) at ground potential surrounds and contains the interrupter(s) and the insulating mediums."

A live tank circuit breaker is defined in the same standard as: "A switching device in which the vessel(s) housing the interrupter(s) is at a potential above ground."

Dead tank circuit breakers are the preferred choice in the US and in most countries that adhere to ANSI published standards. These circuit breakers are said to have the following advantages over the live tank circuit breakers:

1. Multiple low voltage bushing type current transformers can be installed at both, the line side and the load side of the circuit breaker.
2. They have a low, more aesthetic silhouette.
3. Their unitized construction offers a high seismic withstand capability
4. They are shipped factory assembled and adjustments are factory made

In applications where the IEC standards are followed live tank circuit breakers are the norm. The following advantages are generally listed for live tank circuit breakers:

1. Lower cost of circuit breaker (without current transformers)
2. Less mounting space requirements
3. Uses a lesser amount of interrupting fluid

5.1.4 Circuit Breakers Type by Interrupting Mediums

In the evolutionary process of the circuit breaker technology, the main factors that have dictated the overall design parameters of the device are, the interrupting and insulating medium that is used, together with the methods that are utilized to achieve the proper interaction between the interrupting medium and the electric arc.

The choice of air and oil, as the interrupting mediums, was made at the turn of the century and it is remarkable how well and how reliably these mediums have served the industry.

Two newer technologies, one using vacuum and the other sulfurhexafluoride (SF_6) gas as the interrupting medium, made their appearance at about the same time in the late 1950's, and are now what is considered to be the new generation of circuit breakers.

Although vacuum and SF_6 constitute today's leading technologies, a discussion of air and oil circuit breakers is provided because many of these devices are still in service, moreover, because many of the requirements specified in some of the existing standards were based on the operating characteristics of those circuit breakers.

Figure 5.3 Dead tank type circuit breaker (grounded enclosure).

Figure 5.4 Live tank type circuit breaker (ungrounded enclosure).

5.2 Air Magnetic Circuit Breakers

A plain break knife switch operated in free air, under normal atmospheric conditions, is one of the earliest versions known of a circuit breaker. However, this simple device had a very limited capacity in terms of voltage and of interrupting current.

The interrupting capability limitations of such a switch spurred investigations that led to the development of improved designs. Those improvements involved the inclusion of a number of components whose function was to enhance the cooling of the arc.

The most significant advancement was the development of the arc chute, which is a box like component device that contains a number of either metallic or insulated plates. Additionally, in most of the designs, when intended for medium voltage applications, the arc chute includes a magnetic blow-out coil

It is a well-known fact that air at atmospheric pressure has a relatively low dielectric strength, it is also known that in still air nothing accelerates the process of recombination and therefore the time constant for deionization is fairly large.

It should be recalled that a low time constant is highly desirable, in fact is indispensable for high voltage devices; However, a high time constant is acceptable in low and at the lower end of the medium voltage devices, where it offers the advantages of substantially reducing the possibility of generating switching overvoltages, and in many cases of modifying the inherent TRV of the system. This influence may be such that for all practical purposes this type of breakers can be considered to be insensitive to TRV problems.

Since current interruption across an air break is based only upon the natural deionization process that takes place in the air surrounding the arc; then, in order to improve the interrupting capability it is necessary to enhance the deionization process by means of some appropriate external cooling method.

We should recall that in order to maintain the ionization of the gas, when the arc is effectively cooled, the magnitude of the arc voltage must increase. What this means is simply that as the arc cools, the cooling effectively increases the deionization of the arc space, which in turn increases the arc resistance. As a consequence of the increase in resistance, the short circuit current and the phase angle are reduced and thus, the likelihood of a successful interruption is significantly enhanced.

In an air circuit breaker, increasing the resistance of the arc in effect increases the arc voltage Thus, to effectively increase the arc voltage any of the following means can be used.

1. increase the length of the arc, which increases the voltage drop across the positive column of the arc.

2. split the arc into a number of shorter arcs connected in series. What this does is that instead of having a single cathode and anode at the ends of the single arc column there are now a multiplicity of cathode and anode regions, which have additive voltage drops. Although the short arcs reduce the voltage of each individual positive column, the summation of all the voltage drops is usually greater than that of a single column; furthermore, if the number of arcs is large enough so that the summation of these voltage drops is greater than the system voltage a quick extinction of the arc is possible.

3. Constricting the arc, by constraining it between very narrow channels. This in effect reduces the cross section of the arc column and thus increases the arc voltage.

With both of the last two suggested methods there is an added benefit, which is the additional cooling of the arc as the result of the high energy storage capacity provided by the arc chute plates that are housed inside of the arc chute itself.

5.2.1 Arc Chute Type Circuit Breakers

An arc chute can be described as a box shaped structure made of insulating materials. Each arc chute surrounds a single pole of the circuit breaker independently, and it provides structural support for a set of arc plates and in some cases, when so equipped, it houses a built in magnetic blow-out coil

Basically there are two types of arc chutes where each type is characterized primarily by the material of the arc plates that are used. Some arc plates are made of soft steel and in some cases are nickel plated. In this type of arc chute the arc is initially guided inside the plates by means of arc runners, which are simply a pair of modified arc horns. Subsequently the arc moves deeper into the arc chute due to the forces produced by the current loop and the pressure of the heated gases..

To enhance and to control the motion of the arc, vertical slots are cut into the plates. The geometrical pattern of these slots varies among circuit breaker manufacturers, and although there may be some similarities in the plate designs, each manufacturer generally has a unique design of its own.

When the arc comes in contact with the metal plates it divides into a number of shorter arcs that burn across a set of adjacent plates. The voltage drop that is observed across each of these short arcs is usually about 30 to 40 volts. The majority of this voltage being due to the cathode and anode drop of each arc. The voltage drop of the positive column depends in the plate spacing, which in turn, determines the length of the arc's positive column. A schematic representation of this type of arc chute, which is used almost exclusively in low voltage applications, is shown in figure 5.5.

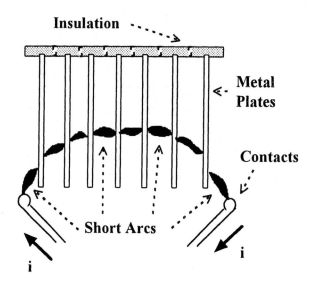

Figure 5.5 Outline of a plain arc chute used in low voltage circuit breakers.

A second type of arc chute is one that is generally synonymous with magnetic blow-out assist and which is used in circuit breakers intended for applications at medium voltages of up to 15 kV and for interrupting symmetrical fault currents of up to 50 kA. This type of arc chute almost invariably uses insulated arcing plates that are made of a variety of ceramic materials such as zirconium oxide or aluminum oxide. An example of this type of circuit breaker is shown in figure 5.6.

With this particular type of arc chute the cooling of the arc and its final quenching is effected by a combination of processes. First, the arc is elongated as it is forced to travel upwards and through a tortuous path that is dictated by the geometry and the location of the insulating plates and their slits.

Simultaneously the arc is constricted as it travels through the slots in the arc plates and as the arc fills the narrow space between the plates. Finally, when the arc gets in contact with the walls of the insulating plates the arc is cooled by diffusion to these walls.

The diffusive cooling is strongly dependent upon the spacing of the plates, this dependence has been shown by G. Frind [5] and his results are illustrated in figure 5.7.

Since the arc behaves like a flexible and stretchable conductor, it is possible to drive the arc upwards forcing it into the spaces between the arcing plates,

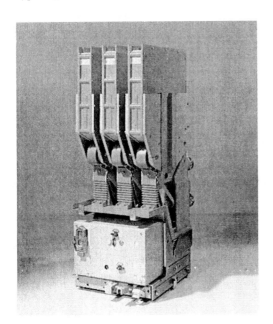

Figure 5.6 Typical 15 kV Air Magnetic circuit breaker.

and thus rather effectively increasing the length of the arc and its resistance.

Motion of the arc is forced by the action the magnetic field produced by a coil that is generally found embedded into the external supporting plates of the arc chute. In figures 5.8 and 5.9 a complete arc chute assembly and a coil that is to be potted are shown.

The coil, which is not a part of the conducting circuit during normal continuos operation, is connected to the ends of an arcing gap as shown schematically in figure 5.10 (a). When the circuit breaker begins to open, the current transfers from the main contacts to the arcing contacts where, upon their separation, the arc is initiated (figure 5.10 (b)) As the contacts continue to increase their separation, the arc is forced into the arc runners where the coil is connected, and in so doing the coil is inserted into the circuit (figure 5.10 (c)) The magnetic field created by the coil will now except a force upon the arc that tends to move the arc up deeper inside the arc chute, figure 5.10 (d).

The heating of the arc plates and their air spaces results in the emission of large amounts of gases and vapors, that must be exhausted through the opening at the top of the arc chute.

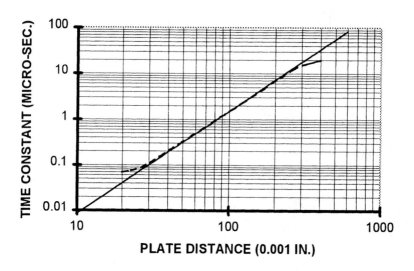

Figure 5.7 Recovery time constant in air as a function of the spacing between the arc chute plates.

The mixture of gases and vapor is prevented from flowing back and into the contacts by the magnetic pressure that results from the interaction of the arc current and the magnetic field. As long as the forces produced by the gas are lower than the magnetic forces, the flow of the gases will be away from the contacts.

An important requirement for the connection of the coil is to maintain the proper polarity relationships so the arc is driven upwards and into the interrupting chamber. It is also important to have a phase lag between the magnetic flux and the current being interrupted so that at current zero there is still a force being exerted on the extinguishing arc.

Because at low current levels the magnetic force is relatively weak most air magnetic circuit breakers include some form of a puffer that blows a small stream of air into the arc, as the arcing contacts separate, to help drive the arc upwards and into the plates.

In most designs, to avoid the possibility of releasing hot, partially ionized gases which may cause secondary flashovers, a flat horizontal stack of metal plates is placed at the exhaust port of the arc chute.

Finally, it should be noted that even though this is one of the oldest techniques of current interruption and that in spite of all the theoretical and experimental knowledge that has been gained over the years, the design of arc chutes remains very much as an art. Theoretical evaluation of a design is very difficult and the designer still has to rely primarily on experimentation.

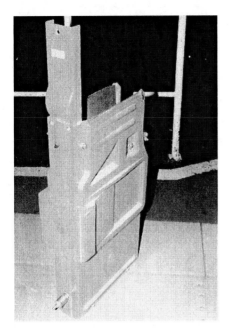

Figure 5.8 Complete assembly of a 15 kV arc chute.

Figure 5.9 Side plate of an arc chute showing the blow-out coil and its assembly location.

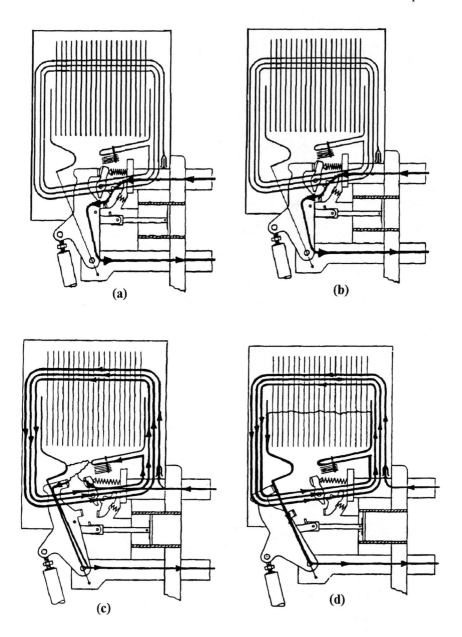

Figure 5.10 Blow-out coil insertion sequence of an air magnetic circuit breaker:
(a) circuit breaker closed, coil by-passed and (b) circuit breaker during arcing period
coil inserted.

5.2.2 Air Magnetic Circuit Breakers: Typical Applications

One important characteristic of air magnetic circuit breakers is that their interrupting capability is greatly influenced by the magnitude of the system voltage. It has been demonstrated that the interrupting current capability increases as the voltage decreases and consequently these breakers can be referred as being a voltage controlled interrupter. This characteristic, as it will be shown in a later chapter, is reflected in some existing performance standards where within some specific limits the interrupting current capability is given by:

$$\frac{V_{Max}}{V_{Min}} = \frac{I_{Max}}{I_{Min}}$$

Because of the high arc resistance that is characteristically exhibited by these circuit breakers, they are capable of modifying the normal wave form of the fault current to the point that they may even advance the occurrence of a current zero. This represents a significant characteristic of these breakers specially for applications in circuits where the fault current asymmetry exceeds 100% and where there may not be currents zeroes for several cycles. This high asymmetry condition is common in applications related to the protection of large generators.

Among some of the significant disadvantages of these circuit breakers when compared to modern type breakers of the same ratings are their size and their cost. Other disadvantages include; short interrupting contact life which is due to the high energy levels that are seen by the interrupter, their need for a relatively high energy operating mechanism and the risks that are posed by the hot gases when they are released into the switchgear compartment following the interruption of a short circuit current.

5.3 Air Blast Circuit Breakers

Although there was a patent issued in 1927, air blast circuit breakers were first used commercially sometime around the year 1940, and for over five decades this technology has proved to be quite successful.

Air blast circuit breakers have been applied throughout the complete high voltage range, and until the advent of SF_6 circuit breakers, they totally dominated the higher end of the transmission voltage class. In fact, at one time they were the only type of circuit breakers available for applications at voltages higher than 345 kV.

In reality, air blast circuit breakers should be identified as a specific type of the more generic class of gas blast circuit breakers because air is not necessarily the only gas that can be used to extinguish the arc, other gases such as nitrogen, carbon dioxide, hydrogen, freon and of course SF_6 can be used. Furthermore, it is well known, and as is generally agreed, the interrupting process is the same for all gas blast circuit breakers and most of the differences in performance observed between air blast and SF6 circuit breakers are the result of the variations in the cooling capabilities, and therefore in the deionization time constant of each of the gases. For this reason the detailed treatment of the interrupting process will be postponed to later in this chapter, when describing the more modern SF$_6$ technology.

Because there are some differences in the basic designs of air blast interrupters, and because the newer concepts for gas blast interrupters have evolved from the knowledge gained with the air blast circuit breaker, there are a number of subjects which need to be addressed in this section if nothing else, to provide a historical frame of reference.

In all of the designs of air blast circuit breakers the interrupting process is initiated by establishing the arc between two receding contacts and by, simultaneously with the initiation of the arc, opening a pneumatic valve which produces a blast of high pressure air that sweeps the arc column subjecting it to the intense cooling effects of the air flow.

5.3.1 Blast Direction and Nozzle Types

Depending upon the direction of the air flow in relation to the arc column [6] there are, as shown in figure 5.11 (a), (b) and (c), three basic types of blast orientations.

1. axial blast
2. radial blast, and
3. cross blast

From the three blast types, the axial or the radial type are generally preferred for extra high voltage applications, while the cross blast principle has been used for applications involving medium voltage and very high interrupting currents.

To effectively cool the arc the gas flow in an axial blast interrupter must be properly directed towards the location of the arc. Effective control of the gas flow is achieved by using a D'Laval type of a converging-diverging nozzle.

These nozzles can be designed either as insulating, or as metallic or conducting nozzles. Additionally and depending in the direction of flow for the exhaust gas each of the nozzles in these two groups, can be either what is called a single or a double flow nozzle.

A conducting single flow nozzle, as shown in figure 5.12, is one where the main stationary contact assembly serves a dual purpose. It carries the continuous current when the circuit breaker is closed and as the circuit breaker opens the arc is initiated across one of its edges and its corresponding mating moving contact.   After the gas flow is established, the pneumatic force exerted by the

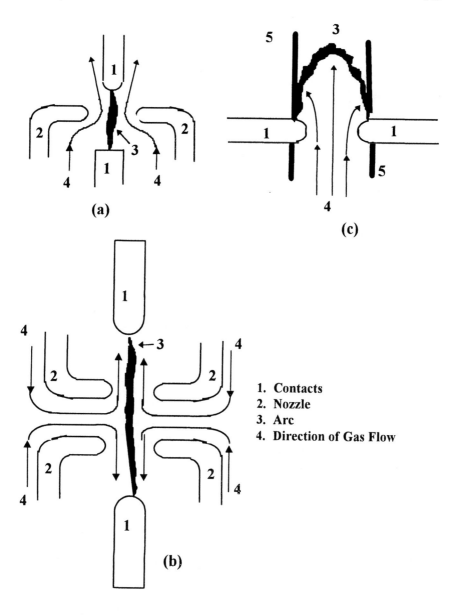

Figure 5.11 Air blast direction: (a) axial direction, (b) radial direction and (c) cross blast or transverse direction.

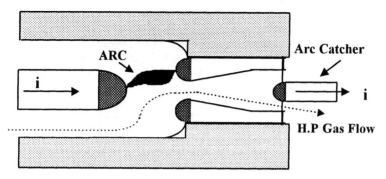

Figure 5.12 Outline of a conducting single flow nozzle.

gas on the arc effectively transfer the arc to a stationary arc terminal, or arc catcher that is disposed longitudinally at the center of the nozzle.

It is easily observed that with this design the arc length can be increased considerably and at a faster rate than that which is possible with an insulating nozzle where the arc is initiated directly across a pair of receding arcing contacts. Under these conditions the time needed for the arc to reach its final length is dependent upon the final contact gap and consequently on the opening speed of the breaker contacts that is normally in the range of 3 to 6 meters per second (10 to 20 feet per second).

An axial insulating nozzle is geometrically similar to the conducting nozzle as shown in figure 5.13 and as its name implies, the insulating nozzle is made of an insulating material. The material of choice is generally teflon, either as a pure compound or with some type of filler material. Fillers are used to reduce the rate of erosion of the throat of the nozzle.

It should be noted that once the arc is properly attached to the intended arcing contacts, the gas flow characteristics, and the interaction between the gas and the arc are the same for both types of nozzles.

The cross blast design is among one of the earliest concepts that was used on air circuit breakers. As shown schematically in figure 5.11 (c) the arc is initiated across a pair of contacts and is subjected to a stream of air that flows perpendicular to the axis of the arc column. It was contended that a considerable amount of heat could be removed from the arc since the whole length of the arc is in contact with the air flow. However, this is not the case, mainly because the core of the arc has a lower density than the surrounding air and therefore at the central part of the arc column there is very little motion between the arc and the gas. Nevertheless, at the regions lying alongside the contacts where the roots of the arc are being elongated the gas flows in an axial direction and a substantial

Insulating Nozzle

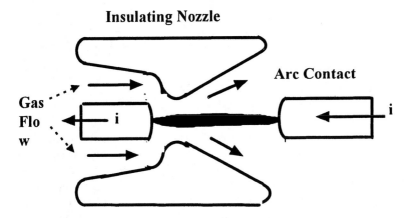

Figure 5.13 Outline of an axial insulating nozzle.

amount of cooling can be achieved. Most of the circuit breakers that were made using a cross blast were applied only at medium voltages and high currents.

5.3.2 Series Connection of Interrupters

Without a doubt, a single break interrupter is the simplest and most economical solution. However, significant improvements in the interrupting capacity of a circuit breaker can be achieved by connecting a number of interrupters in series.

It is easy to see that by connecting a number of interrupters in series, the recovery voltage, at least in theory, is equally divided across each interrupter, it also can be seen that the number of deionizing chambers is increased and thus the energy balancing process is increased proportionally to the number of interrupters connected in series.

One of the main difficulties encountered with this type of application is to ensure that during the transient period of the interruption process each of the interrupters operates under the exact same conditions of each other. This sameness requirement applies to both, the aerodynamic and the electrical conditions. From the aerodynamic point of view the flow conditions must be maintained for each interrupter. This generally requires the use of individual blast valves for each interrupter. It also requires that the lines connecting each interrupting chamber are properly balanced to avoid pressure drops that may affect the gas flow.

Electrically, the restriking voltages and consequently the recovery voltages, must divide evenly across each set of contacts. However, in actual practice this does not happen and an uneven distribution of voltage occurs due to the

unbalanced inherent capacitance that exists across the interrupting device itself, and between the line and the grounded parts of the circuit breaker. To
improve the voltage distribution across the gaps it is a common practice to use
grading capacitors or resistors connected between the live parts and ground.

5.3.3 Basic Interrupter Arrangements

Medium voltage air blast circuit breakers were normally dead tank designs.
Air circuit breakers intended for applications at systems voltages greater than
72.5 kV were almost without exception of the live tank design type.

In some of the earlier designs the blast valve was located inside of the high
pressure storage tank, at ground level, while the interrupters were housed at the
end of insulating columns. The tripping operation was initiated by opening the
blast valve which in turn momentarily pressurized the interrupter causing the
contacts to move. The blast valve was closed later, in about 100 milliseconds,
and as the pressure inside of the interrupters was decreased the breaker contacts reclose.

To maintain system isolation in the open position of the circuit breaker
there was a built-in plain air break isolating switch that also served as a dis-
connecting switch for the grading capacitors, or resistors. The isolating switch
was timed to open in about 40 to 50 milliseconds after the opening time of the
main contacts.

The disadvantages associated with this type of design were the larger gap
length required by the isolator at the higher system voltages. As it can be expected the greater gaps required additional operating time and therefore fast
reclosing times were difficult to achieve. Furthermore, since the exhaust was
open to the atmosphere the air consumption was relatively large.

A number of improvements where made focused primarily with the objective of removing the need for the air isolating switch. Many different arrangements of blast and exhaust valves where used until the present design in
which the interrupters are maintained fully pressurized at all times was
adopted. In this version tripping of the circuit breaker is executed by first
opening the exhaust valves and then sequentially opening the contacts. After a
few milliseconds the exhaust valves are closed while the contacts still are in
the open position. For closing the contacts are depressurized while the valves
are held closed.

With this design, the air consumption was substantially reduced and what is
more important, since the interrupting chambers were held at the maximum
pressure at all times, the breaking and the withstand capacity of the interrupter
was optimized. One last advantage that should be mentioned is that since the
air consumption was reduced so was the operating noise level of the interrupter. This is significant because air blast circuit breakers any notorious for
their high operating noise level.

5.3.4 Parameters Influencing Air Blast Circuit Breaker Performance

There are many factors that influence the performance of a gas blast circuit breaker. However, from all those factors there are some that can be easily measured such as the operating pressure, the nozzle diameter and the interrupting current.

These parameters have been used to establish relationships [7], [8], that are related to the voltage recovery capability of the interrupter in the thermal region. These relationships are included in their graphic form in figures 5.14, 5.15, and 5.16.

The significance of these relationships is not in the absolute values that are being presented because, depending on the specific design of the nozzle and in the overall efficiency of the interrupter, the magnitude of the variables change; however, the slope of the curves and therefore the exponent of the corresponding variables remains constant thus indicating a performance trend and giving a point of reference for comparison between interrupter designs. Furthermore and as it will be seen later there is a great deal of similarity between these curves and those obtained for SF_6 interrupters.

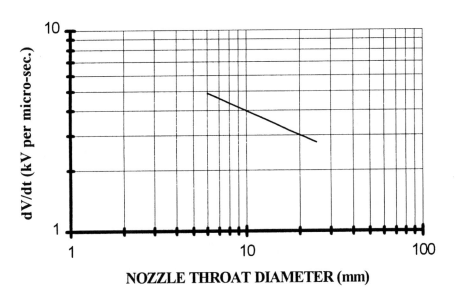

Figure 5.14 Voltage recovery capability in the thermal region as a function of nozzle throat diameter for an air blast interrupter.

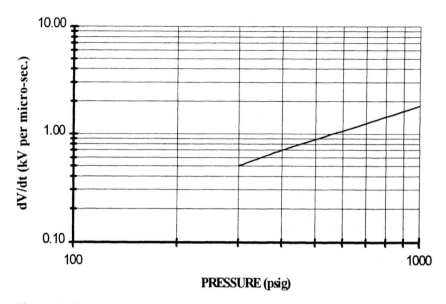

Figure 5.15 Voltage recovery capability in the thermal region as a function of pressure for an air blast interrupter.

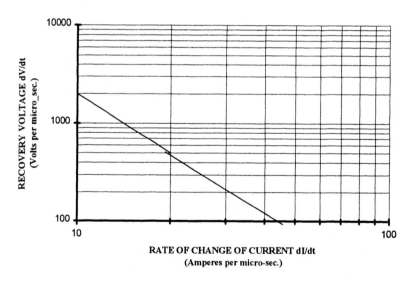

Figure 5.16 Voltage recovery capability in the thermal region as a function of rate of change of current for an air blast interrupter.

5.4 Oil Circuit Breakers

From a historical perspective, the oil circuit breaker is the first design of a breaker for high power applications. It predates the air blast type by several decades.

One of the first designs of an oil circuit breaker on record in the US, is the one shown in figure 5.17. This circuit breaker was designed and built by J. N. Kelman in 1901. The breaker was installed on a 40 kV system that was capable of delivering a maximum short circuit current of 200 to 300 amperes. Records indicate that the circuit breaker was in service from April 1902 until March of 1903, when following a number of circuit interruptions, at short time intervals, blazing oil was spewed over on the surrounding woodwork, starting a fire which eventually spread to the power house [9].

The design of this circuit breaker was extremely simple. It consisted of two wooden barrels filled with a combination of water and oil. The contacts consisted of two vertical blades connected at the top and arranged so that they would drop into the stationary contacts to close the circuit.

Figure 5.17 Oil circuit breaker built in 1901 by Kelman (reference 9).

From these relatively humble beginnings the oil circuit breaker was refined and improved but, throughout all these mutations it maintained its characteristic simplicity of construction and its capability for interrupting large currents.

Oil circuit breakers where widely used and presently there are many still in service. However, they have suffered the same fate as did the air blast circuit breaker, they have been made obsolete by the new SF_6 technology.

5.4.1 Properties of Insulating Oil

The type of oil that has been used in virtually all oil circuit breakers is one where naphthenic base petroleum oils have been carefully refined to avoid sludge or corrosion that may be produced by sulfur or other contaminants.

The resulting insulating oil is identified as type 10-C transformer oil. It is characterized by an excellent dielectric strength, by a good thermal conductivity (2.7×10^{-4} cal/sec cm °C) and by a high thermal capacity (0.44 cal/g °C).

Some designs of oil circuit breakers take advantage of the excellent dielectric withstand capabilities of oil and use the oil not only as interrupting medium but also as insulation within the live parts of the circuit breaker and to ground.

Insulating oil at standard atmospheric conditions, and for a given contact gap, is far superior than air or SF_6 under the same conditions. However, oil can be degraded by small quantities of water, and by carbon deposits that are the result of the carbonization of the oil. The carbonization takes place due to the contact of the oil with the electric arc.

The purity of the oil usually can be judged by its clarity and transparency. Fresh oil has a clear amber color, while contaminated oil is darken and there are some black deposits that show signs of carbonization. The condition of the oil normally is evaluated by testing for its withstand capability. The tests are made using a spherical spark gap with two spheres 20 mm in diameter and at a gap of 3 mm.

Fresh oil should have a dielectric capability greater than 35 kV. For used oil it is generally recommended that this capability be no less than 15 kV.

5.4.2 Current Interruption in Oil

At the time when the oil circuit breaker was invented no one knew that arcs drawn in oil formed a bubble containing mainly hydrogen and that arcs burning in a hydrogen atmosphere tend to be extinguished more readily than arcs burning in other types of gases. The choice of oil was then indeed a fortuitous choice that has worked very well over the years.

When an arc is drawn in oil the contacting oil surfaces are rapidly vaporized due to the high temperature of the arc, which as we already know is in the range of 5,000 to 15,000°K. The vaporized gas then forms a gas bubble, which totally surrounds the arc.

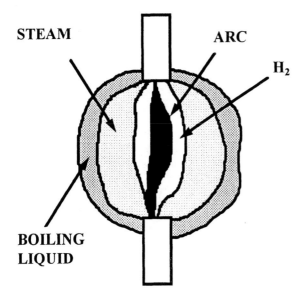

STEAM   **ARC**

                        H₂

BOILING
LIQUID

Figure 5.18 Gas bubble produced by an arc that is surrounded by oil.

It has been observed that the approximate composition of this bubble is 60 to 80% hydrogen, 20% acetylene (C_2H_2) and the remainder consists of smaller proportions of methane and other gases.

Within the gas bubble, shown in figure 5.18, there are at three easily identifiable zones. In the innermost zone, which contains the dissociated gases and is the one in direct contact with the arc, it has been observed that the temperature drops to between 500 to 800(K. This gaseous zone is surrounded by a vapor zone where the vapor is superheated in its inside layers and is saturated at the outside layers. The final identifiable zone is one of boiling liquid where at the outside boundary the temperature of the liquid is practically equal to the relative ambient temperature.

Considering that the arc in oil circuit breakers is burning in a gaseous atmosphere it would be proper to assume that the theories of interruption developed for gas breakers are also applicable to the oil breaker. This assumption is been proven to be correct and therefore the performance of both, gas blast circuit breakers, as well as oil circuit breakers, can be evaluated by applying the theories of arc interruption that were presented in chapter 1.

It has been demonstrated that hydrogen is probably the ideal gas for interruption, but the complications and cost of a gas recovery system make its application impractical.

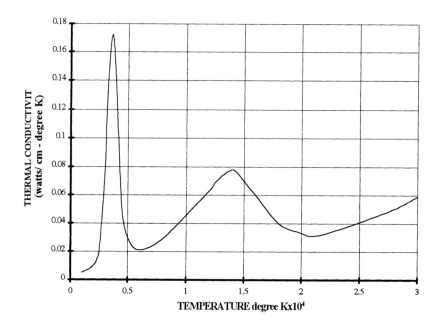

Figure 5.19 Thermal conductivity of hydrogen.

Even though that comparatively speaking, the dielectric strength of hydrogen is not particularly high, its reignition voltage is 5 to 10 times higher than that of air. Hydrogen also has a very high thermal conductivity that is faster during the period of gas dissociation, as shown in figure 5.19, which results in a more rapid cooling and deionizing of the arc

5.4.3 Types of Oil Circuit Breakers

In the earlier designs of oil circuit breakers the interrupters consisted of only a plain break and no consideration was given to include special devices to contain the arc or to enhance the arc extinguishing process. In those early designs the arc was merely confined within the walls of a rather large oil tank and deionization was accomplished by (a) elongation of the arc, (b) by the increased pressure produced by the heating of the oil in the arc region and (c) by the natural turbulence that is set by the heated oil. This plain break circuit breaker concept is illustrated in figure 5.20.

To attain a successful interruption, under these conditions, it is necessary to develop a comparatively long arc. However, long arcs are difficult to control, and in most cases this leads to long periods of arcing.

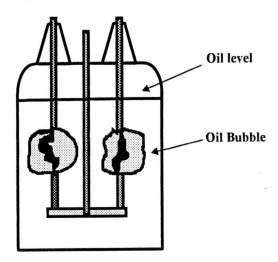

Oil level

Oil Bubble

Figure 5.20 Outline of a plain break oil circuit breaker.

The random combinations of long arcs, which translate into high arc voltages, accompanied by long arcing times make unpredictable the amount of arc energy that has to be handled by the breaker. This unpredictability presents a problem because it is not possible to design a device that can handle such a wide and non well-defined range of energy.

Plain break oil circuit breakers where generally limited on their application to 15 kV systems and maximum fault currents of only about 200 amperes. Moreover, these circuit breakers were good only in those situations where the rate of rise of the recovery voltage was low.

The development of the explosion chamber, or interrupting pot, constituted a significant breakthrough for oil circuit breakers. It led to the designs of the so called "suicide breakers". Basically the only major change made on the plain breaker design was the addition of the explosion pot, which is a cylindrical container fabricated from a mechanically strong insulating material. This cylindrical chamber is mounted in such a way as to fully enclose the contact structure. At the bottom of the chamber there is an orifice through which the moving contact rod is inserted.

The arc as before is drawn across the contacts, but now it is contained inside the interrupting pot and thus the hydrogen bubble is also contained inside the chamber. As the contacts continue to move and whenever the moving contact rod separates itself from the orifice at the bottom of the chamber an

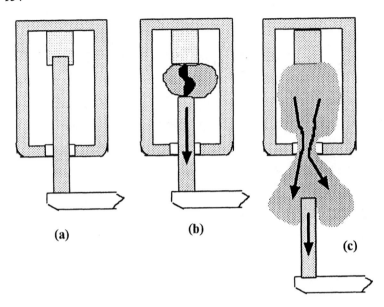

Figure 5.21 Outline of an explosion chamber type of oil interrupter. (a) Contacts closed, (b) arc is initiated as contacts move, (c) gas escapes through interrupter pot opening.

exit similar to a nozzle becomes available for exhausting the hydrogen that is trapped inside the interrupting chamber. A schematic drawing of this design is included in figure 5.21.

One of the disadvantages of this design is its sensitivity to the point on the current wave where the moving contact rod is separated from the interrupter chamber. If the first current zero occurs too early before the contact leaves the bottom orifice then the interrupter must wait for the next current zero which may come a relatively long time after the contact had left the pot and consequently when the pressure inside the pot has decayed to an ineffective value due to the venting through the bottom orifice.

Another drawback of this interrupter chamber is its dependency on current magnitude. At high values of current the corresponding generated pressure is high and may even reach levels that would result in the destruction of the chamber. Sometimes the high pressure has a beneficial quasi-balancing effect because the high pressure tends to reduce the arc length and the interrupting time, thus, decreasing the arc energy input. However, with lower values of current, the opposite occurs, the generated pressures are low and the arcing times

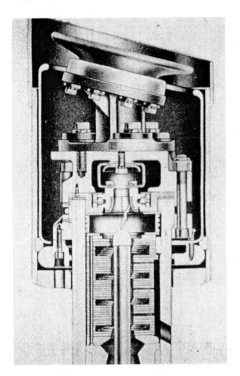

Figure 5.22 Cross baffle interrupter chamber.

increase until a certain critical range of current, reached where it is difficult to achieve interruption. This current level is commonly identified as the "critical current."

Among the alternatives developed to overcome these limitations are the inclusion of pressure relief devices to limit the pressure due to the high currents. For the low current problem, the impulse breaker was developed. This design concept provides a piston pump intended to squirt oil into the contacts at the precise time when interruption is taking place.

To reduce the sensitivity to the contact position at current zero the cross baffle interrupter chamber design was created. This design rapidly gained popularity and it became the preferred design for all the later vintage oil circuit breakers. A typical interrupting chamber of this type is shown in figure 5.22.

The design consists of a number of specially designed insulated plates that are stacked together to form a passage for the arc that is alternately restricted

Figure 5.23 Oil breaker interrupting chamber showing lateral vents.

and then laterally vented, as shown in figure 5.23. This design permits the lateral venting of the pressure generated inside of the chamber. This arrangement subjects the arc to a continuous strong cross flow which has proven to be beneficial for extinguishing the arc.

Further developments of the interrupting chambers led to some designs that incorporated cross blast patterns, while others included what is known as compensating chambers where an intermediate contact is used to establish the arc sequentially. The first contact draws the arc in an upper chamber which preheats the oil prior to opening the second contact.

A typical relationship between the arcing time as a function of the interrupted current and as a function of the system voltage was established by F. Kesselring [10] and is shown in figure 5.24 (a) and (b).

5.4.4 Bulk Oil Circuit Breakers

The main distinguishing characteristic of bulk oil breaker types is the fact that these circuit breakers use the oil not only as the interrupting medium but also as the primary means to provide electrical insulation.

The original plain break oil circuit breakers obviously belonged to the bulk oil circuit breaker type. Later, when the newly developed interrupting chambers where fitted to the existing plain break circuit breakers with no significant

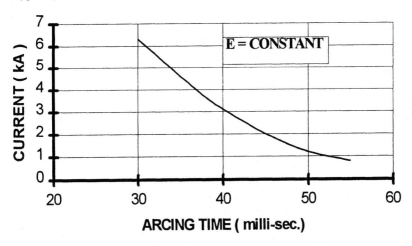

Figure 5.24 (a) Oil breaker arcing time as function of current at constant voltage.

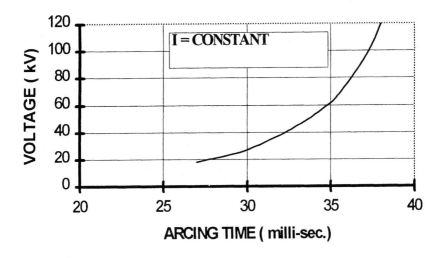

Figure 5.24 (b) Oil breaker arcing time as function of voltage at constant current.

modifications being made specially to the oil tank, and because of the good acceptance of this type of design, specially in the US; the bulk oil type concept was simply continued to be fabricated.

Figure 5.25 (a) 15 kV single tank oil circuit breaker .

In many cases, at voltages that generally extended up to 72.5 kV all three poles were enclosed into a single tank of oil, However a number of breakers in the medium voltage range had three independent tanks, as did those circuit breakers with voltage ratings greater than 145 kV. The three poles were gang operated by a single operating mechanism. The single and the multiple tank circuit breaker designs are shown in figure 5.25 (a) and (b).

Figure 5.25 (b) 230 kV multitank oil circuit breaker.

To meet the insulating needs of the equipment, and depending on the magnitude of the application voltage, adequate distances must be provided between the live parts of the device and the grounded tank containing the insulating oil. Consequently this type of design required large tanks and large volumes of oil, for example for a 145 kV circuit breaker approximately 12,000 liters, or about 3,000 gallons of oil were required, and for a 230 kV circuit beaker the volume was increased to 50,000 liters, or approximately 13,000 gallons.

Not only the size of the breakers was very large but also the foundation pads where the breakers were mounted had to be big and quite strong. In order to withstand the impulse forces developed during interruption it is usually required that the pad be capable of supporting a force equal to up to 4 times the weight of the circuit breaker including the weight of the oil. This in the case of a 245 kV circuit breaker amounted to a force of about 50 tons.

5.4.5 Minimum Oil Circuit Breakers

Primarily in Europe, because of the need to reduce space requirements and the scarcity and high cost of oil, a new circuit breaker which uses very small volumes of oil was developed. This new circuit breaker is the one known by any of the following names; minimum oil, low oil content, or oil poor circuit breaker.

The main difference between the minimum oil and the bulk oil circuit breakers is that minimum oil breakers use oil only for the interrupting function while a solid insulating material is used for dielectric purposes, as opposed to bulk oil breakers where oil serves both purposes.

In minimum oil circuit breakers a small oil filled, arc interrupting chamber is supported within hollow insulators. These insulators are generally fabricated from reinforced fiber glass for medium voltage applications and of porcelain for the higher voltages.

Figure 5.26 Typical 15 kV minimum oil circuit breaker.

The use of insulating supports effectively qualify this design as a live tank breaker. By separating the live parts from ground by means of the insulating support the volume of oil required is greatly decreased as it can be seen in figure 5.26 where a typical 15 kV low oil breaker is shown.

5.5 Sulfurhexafluoride

Considering the fact that oil and air blast circuit breakers have been around for almost one hundred years; sulfurhexafluoride circuit breakers are a relatively newcomer, having been commercially introduced in 1956.

Although SF_6 was discovered in 1900 by Henry Moissan [11] The first reports of investigations made exploring the use of SF_6 as an arc quenching medium was published in 1953 by T. E. Browne, A. P. Strom and H. J. Lingal [12]. These investigators made a comparison of the interrupting capabilities of air and SF_6 using a plain break interrupter.

The published results, showing the superiority of SF_6 were simply astounding. As it can be seen in figure 5.27 SF_6 was 100 times better than air. In the same report it was shown that the addition of even moderate rates of gas flow increased the interrupting capability by a factor of 30.

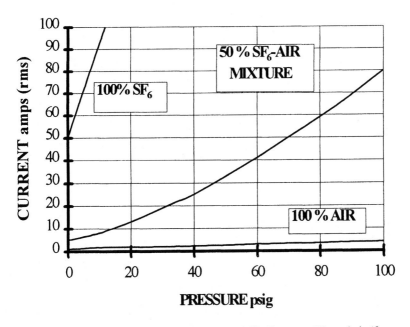

Figure 5.27 Comparison of interrupting capability between SF_6 and air (from ref. 12).

Chapter 5

SF$_6$ circuit breakers, in their relatively short existence already have come to completely dominate the high voltage circuit breaker market and in the process they have made obsolete the air blast and oil technologies. Almost without exception SF$_6$ circuit breakers are used in all applications involving system voltages anywhere in the range of 72.5 kV to 800 kV.

In medium voltage applications, from 3 kV and up to about 20 kV, SF$_6$ has found a worthy adversary in another newcomer the vacuum circuit breaker. Presently neither technology has become the dominant one, although there are strong indications that for medium voltage applications vacuum may be gaining an edge.

5.5.1 Properties of SF$_6$

SF$_6$ is a chemically very stable, non-flammable, non-corrosive, non-poisonous, colorless and odorless gas. It has a molecular weight of 146.06 and is one of the heaviest known gases. The high molecular weight and its heavy density limits the sonic velocity of SF6 to 136 meters per second which is about one third that of the sonic velocity of air.

SF$_6$ is an excellent gaseous dielectric which, under similar conditions, has more than twice the dielectric strength of air and at three atmospheres of absolute pressure it has about the same dielectric strength of oil (figure 5.28). Furthermore, it has been found that SF$_6$ retains most of its dielectric properties when mixed even with substantial proportions of air, or nitrogen.

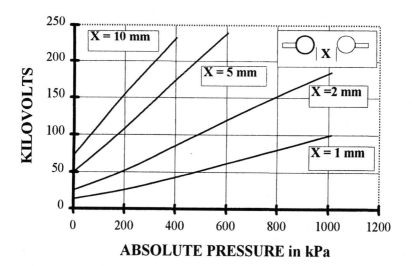

ABSOLUTE PRESSURE in kPa

Figure 5.28 Dielectric strength of SF$_6$ as function of pressure.

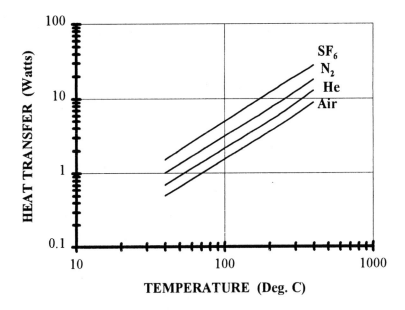

Figure 5.29 Heat transfer characteristics of SF$_6$.

Because of its superior heat transfer capabilities, which are shown in figure 5.29, SF$_6$ is better than air as a convective coolant. It should be noted that while the thermal conductivity of helium is ten times greater than that of SF$_6$, the later has better heat transfer characteristics due to the higher molar heat capacity of SF$_6$ which together with its low gaseous viscosity enables it to transfer heat more effectively.

SF$_6$ is not only a good insulating gas but it is also an efficient electron scavenger due to its electron affinity or electronegativity. This property is primarily responsible for its high electric breakdown strength, but it also promotes the rapid recovery of the dielectric strength around the arc region following the extinction of the arc.

Because of its low dissociation temperature and its high dissociative energy SF$_6$ is an excellent arc quenching medium. Additionally, the outstanding arc extinguishing characteristics of SF$_6$ are also due to the exceptional ability of this gas to recover its dielectric strength very rapidly following a period of arcing, and to its characteristically small time constant which dictates the change of conductance near current zero.

The first characteristic is important for bus terminal faults while the second is essential for the successful interruption of short line faults.

5.5.2 Arc Decomposed By-products

At temperatures above 500(C. SF_6 will begin to dissociate. The process of dissociation can be initiated by exposing SF_6 to a flame, electrical sparking, or an electric arc. During this process the SF_6 molecules will be broken down into sulphur and fluorine ions at a temperature of about 3000(C.

It should be recalled that during the interruption process the core of the arc will reach temperatures well in excess of 10,000(K; However, after the arc is extinguished and the arc region begins to cool down and when the temperature drops below approximately 1,000(C the gas will begin to recombine almost totally, and only a small fraction will react with other substances..

The small amounts of gas that do not recombine react with air, with moisture, with the vaporized electrode metal and with some of the solid materials that are used in the construction of the circuit breaker. These decomposition by-products may be gaseous or solid, but they essentially consist of lower sulfur fluorides, and of metal fluorides of which the most notables are CuF_2, AlF_3, WF_6, CF_4.

Among the secondary sulphur-fluorides compounds that are formed [13] are S_2F_2 and SF_4, but they quickly react with moisture to yield hydrogen fluoride (HF), sulfur dioxide (SO_2) and other more stable oxyfluorides such as thionyl fluoride SOF_2.

The metallic fluorides are usually present in the form of a fine non-conductive dust powder that is deposited on the walls and in the bottom of the breaker enclosure. In the case of copper electrodes the solid substances appear as a milky white powder that acquires some blue tinges when exposed to the air atmosphere due to a reaction which yields a dehydrated salt.

5.5.2.1 Corrosive Effects of SF_6

Sulphur-hexafluoride on its pure and uncontaminated form is a non-reactive gas and consequently there is no possibility for any type of corrosion that may be directly attributable to SF6.

When the by-products of arced SF_6 come in contact with moisture some corrosive electrolytes may be formed. The most commonly used metals generally do not deteriorate and remain very stable. However, phenolic resins, glass, glass reinforced materials and porcelain can be severely affected. Other types of insulating materials such as polyurethane, Teflon (PTEE) and epoxies, either of the bisphenol A or the cycloaliphatic type, are unaffected.

It is therefore very important to take appropriate measures in the selection of materials, and the utilization of protective coatings. Corrosion can also be prevented by the elimination of moisture.

5.5.2.2 By-products Neutralization

The lower fluorides and many of the other by-products are effectively neutralized by soda lime (a 50-50 mixture of NaOH + CaO), by activated alumina (especially, dried Al_2O_3), or by molecular sieves.

The preferred granule size for soda lime or alumina is 8 to 12 mesh, but these do not exclude the possible use of other mesh sizes. The recommended amount to be used is approximately equal to 10 % of the weight of the gas.

Removal of the acidic and gaseous contaminants is accomplished by circulating the gas through filters containing the above described materials. These filters can either be attached to the circuit breaker itself or they may be installed in specially designed but commercially available gas reclamation carts.

If it is desired to neutralize SF_6 which has been subjected to an electric arc, it is recommended that the parts be treated with an alkaline solution of lime $(Ca(OH)_2)$, Sodium Carbonate (Na_2CO_3), or Sodium bicarbonate $(NaHCO_3)$.

5.5.3 SF$_6$ Environmental Considerations

The release of human made materials into the atmosphere has created two major problems. One is the depletion of the stratospheric ozone layer and the other is the global warming or "green house effect."

Ozone Depletion Agent. SF_6 does not contribute to the ozone depletion for two reasons: First because due to the structure of the ultraviolet absorption spectrum of SF_6 the gas can not be activated until it reaches the mesosphere at about 60 kilometers above the earth and this altitude is far above the stratospheric one which is in the range of about 30 to 45 kilometers. The second reason is the fact that SF_6 does not contain chlorine which is the principal ozone destroying agent.

Green house effects Agent. SF_6 has been labeled as the most potent greenhouse gas ever evaluated by the scientists of the Intergovernmental Panel on Climate Change (IPCC) [14] [15].

What makes SF_6 such a potentially powerful contributor to global warming is the fact that SF_6, like all the compounds in the fully fluorinated family, has a super stable molecular structure. This structure makes these compounds to be very long lived, to the extent that within human time frames these gases are indestructible.

SF_6 is a very good absorber of infrared radiation. This heat absorption characteristic combined with its long life (3,200 years) [16] has led scientists to assign an extremely high Global Warming Potential (GWP rating to SF_6.

The GWP rating is a comparative numerical value that is assigned to a compound. The value is arrived at by integrating over a time span the radiative forcing value produced by the release of 1 kg of the gas in question and then dividing this value by the value obtained with a similar procedure with CO_2. Because CO_2 is considered the most common pollutant, it has been selected as the basis of comparison for assigning GWP values to other pollutants.

The radiative forcing, according to its definition, is the change in net irradiance in watts per square meter.

The GWP values for CO_2 and for the most common fully fluorinated compounds integrated over a one hundred years time horizon are given in Table 5.1 (taken from reference [14]).

TABLE 5.1

Global Warming Potential (GWP)for most common FFC's
compared to CO_2

COMPOUND	LIFETIME YEARS	GWP
CO_2	50 --200	1
CF_4	50,000	6,300
C_2F_6	10,000	12,500
SF_6	3,200	24,900
C_6F_{14}	3,200	6,800

Presently the concentration of SF_6 has been reported as being only about 3.2 parts per trillion by volume (pptv). This concentration is relatively low, but it has been observed that it is increasing at a rate of about 8 % per year. This means that if the concentration continues to increase at this rate, in less than 30 years the concentration could be about 50 pptv.

More realistically, assuming a worst case scenario [16], the concentration of 50 pptv is expected to be reached by the year 2100. A more optimistic estimate is 30 pptv. At these concentrations the expected global warming attributable to SF_6 has been calculated as 0.02 and 0.014 °C for the most pessimistic and the most optimistic scenarios respectively. Additional data indicate that the expected global warming due to SF_6 through the year 2010 is about 0.004 °C. In comparison with an increase of 300 parts per million by volume (ppmv), of CO_2, the expected change in the global temperature is 0.8 °C.

It is apparent that based on the estimated emission rates the concentration of SF_6 would be very small [17]. Nevertheless, because of the long life time of SF_6 there is a potential danger, specially if the rate of emissions where to increase rather than to reach a level value. It is therefore essential that all types of release of SF_6 into the atmosphere be eliminated or at least reduced to an absolute minimum. This can be done by strict adherence to careful gas handling procedures and proper sealing for all new product designs.

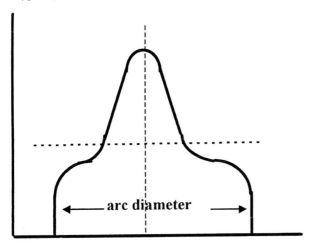

Figure 5.30 Electric arc, radial temperature profile.

5.5.4 Current Interruption in SF$_6$

As we know, the electric arc is a self-sustaining discharge consisting of a plasma that exists in an ionized gaseous atmosphere. We also know that the plasma has an extremely hot core surrounded by an atmosphere of lower temperature gases.

Figure 5.30 represents the typical temperature profile of an arc as a function of its radius, when the arc is being cooled by conduction. The figure shows that there is a relatively thin central region of very high temperature corresponding to the core of the arc. It also shows the existence of a broader, lower temperature region and the transition point between these two regions, where there is a rather sharp increase in temperature.

This characteristic temperature profile simply indicates that the majority of the current is carried by the hottest region of the arc's core which is located close to the central axis, the reason being, as we well know that an increase in temperature corresponds to an increase in electrical conductivity.

Since the arc always tries to maintain its thermal equilibrium, its temperature will automatically adjust itself in relation to the current magnitude. However, once full ionization is attained further increases in current do not lead to increases in temperature. Nevertheless, as the current approaches zero the temperature about the core of the arc begins to drop and consequently the region starts losing its conductivity.

The peak thermal conductivity of SF$_6$, as is seen in figure 5.31, occurs at around 2,000 °K; therefore, near current zero, when rapid cooling is needed for

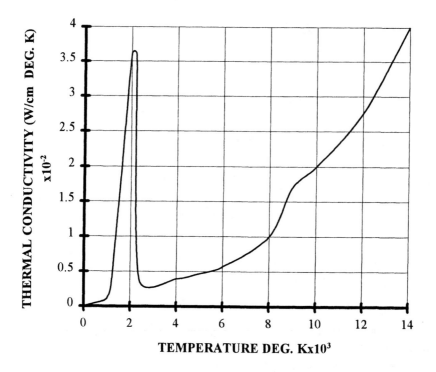

Figure 5.31 Thermal conductivity of SF_6.

interruption, SF_6 is extremely effective because at this temperature electrical conductivity is very low.

At the other side of the spectrum, at high currents the thermal conductivity of SF_6 is not much different from that of other gases and therefore the arc cooling process in that region is about the same regardless of the kind of gas that is being used.

The main difference between interruption in air and in SF_6 is then the temperature at which maximum thermal conductivity takes place. These temperatures are about 6,000 °K for air and 2,000 °K for SF_6.

This difference translates into the fact that SF_6 is capable of cooling much more effectively than air at the lower temperatures and therefore it is capable of withstanding higher recovery voltages sooner. In other words the time constant of SF_6 is considerably shorter than that of air.

The assigned time constant for SF_6 is 0.1 microseconds while for air is greater than 10 microseconds. The significance of this time constant is appreciated when consideration is given to applications where a high rate of TRV is expected, such as in the case of short line faults. Experience indeed has shown that SF_6 can withstand higher recovery rates than air.

5.5.5 Two Pressure SF_6 Circuit Breakers

The first SF_6 circuit breaker rated for application at voltages higher than 230 kV and a current interrupting capability of 25 kA was commercially introduced by Westinghouse in 1959.

The original design of this type of circuit breaker was an adaptation of the air blast and oil circuit breaker designs, and thus the axial blast approach, which was described before when discussing air blast breakers, was used. Naturally, the main difference was, that air had been replaced by SF_6.

The new circuit breakers were generally of the dead tank type. The construction of the tanks, together with their substantial size and strength was quite similar to the tanks used for oil breakers as it can be seen in figure 5.32.

In many cases even the operating mechanisms that had been used for oil circuit breakers where adapted to operate the SF_6 breaker.

The conscious effort made to use some of the ideas from the older technologies is understandable, after all, the industry was accustomed to this type of design and by not deviating radically from that idea made it easier to gain acceptance for the new design.

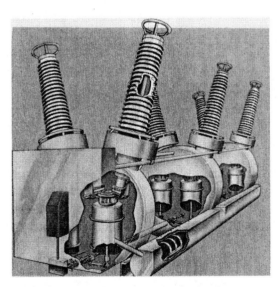

Figure 5.32 Cutaway view of an SF_6 two pressure ITE type GA circuit breaker.

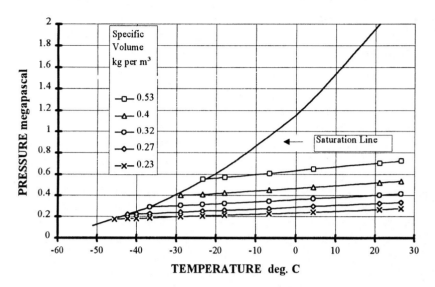

Figure 5.33 Pressure-Temperature variation at constant density for SF₆.

Two pressure circuit breakers were fabricated in either a single or a three tank version, depending mainly in the assigned voltage rating of the device. Smaller high pressure reservoirs were installed next to the low pressure tanks and they were connected to blast valves that operated in synchronism with the contacts. The operating gauge pressures for these circuit breakers were generally around 0.2 MPa for the low side and 1.7 MPa for the high side (30 psig and 245 psig respectively).

The two pressure circuit breaker design prevailed in the US market until the mid-nineteen seventies. At around that time is when the single pressure breakers began to match the interrupting capabilities of the two pressure circuit breakers and thus they became a viable alternative.

Cited among the advantages of the two pressure circuit breaker was that it required a lower operating energy mechanism when compared to the one that is used on single pressure breaker designs. However, in the context of total energy requirements, one must take into account the energy that is spent in compressing the gas for storage and also the additional energy that is required to prevent liquefaction of the SF₆ at low ambient temperatures.

The liquefaction problem represents the main disadvantage of the two pressure breaker. As it can be seen in figure 5.33 at 1.7 MPa the gas will begin to liquefy at a temperature of approximately 13 °C. To prevent liquefaction, and the consequent drop in the gas density electric heaters are installed in the high pressure reservoir.

Liquefaction of SF_6 not only lowers the dielectric capabilities of the gas but it can lead to another problem known as moisture pumping [18] which may happen because of the difference in the condensation point between air and SF_6.

The problem begins in the high pressure system when the gas liquefies in a region that is some distance away from the high pressure reservoir. If the temperature is not sufficiently low to cause condensation of whatever amount of moisture was present in that region then only the liquefied gas will flow back into the reservoir leaving the moisture behind

Since in the mean time, the temperature of the gas in the high pressure reservoir is kept above the dew point, then, the warmer gas will flow back into the breaker attempting to maintain the original pressure. Whatever small amount of moisture is present in the gas contained in the reservoir it will then be transported to the region where liquefaction is taking place. As the gas liquefies again, then once more it will leave the moisture behind. This process can continue until the pressure-temperature conditions change. However, during this time, moisture can accumulate significantly at the coldest point of the gas system, thus increasing the total concentration and reducing the dielectric capability.

Other disadvantages noted are; the high volumes of gas needed, the propensity for higher leak rates due to the higher operating pressures and the added complexity that results from the use of the blast valves.

5.5.6 Single Pressure SF_6 Circuit Breakers

Single pressure breakers have been around at least as long as the two pressure breakers, but initially these breakers where limited to applications requiring lower interrupting ratings.

Later investigations and advanced developments provided answers that led to new designs that had greater interrupting capabilities and around the year 1965 high interrupting capacity puffer breakers were introduced in Europe and in the US.

Puffer circuit breakers have been designed as either dead or live tank as illustrated in figures 5.34 and 5.35.

Customarily, single pressure circuit breakers are described as belonging to either the puffer or the self blast family. But, in reality, all single pressure circuit breakers could be thought as being a member of the self blast family because in either type of circuit breaker the increase in pressure that takes place inside of the interrupter is achieved without the aid of external gas compressors.

The most notable difference between these two breaker types is that in puffer breakers the mechanical energy provided by the operating mechanism is used to compress the gas, while self blast breakers use the heat energy that is liberated from the arc to raise the gas pressure.

Figure 5.34 Dead tank puffer type circuit breaker ABB Power T&D.

Figure 5.35 Merlin Gerin live tank puffer circuit breaker.

Puffer Circuit Breakers. The conceptual drawings and the operating sequence of a typical puffer interrupter is shown in figure 5.36 (a), (b), (c), and (d). A unique characteristic of puffer interrupters is that all have a piston and cylinder combination which is assembled as an integral part of the moving contact structure.

Referring to figure 5.36, (a) shows the interrupter in the closed position, where the volume (V) can be seen. During an opening operation the main contacts separate first, followed by the arcing contacts, figure 5.36 (b). The motion of the contacts decrease the dead volume (V), and thus compress the gas contained within that volume.

As the contacts continue to separate the volume is further compressed, and at the instant when the arcing contact leaves the throat of the nozzle the flow of gas along the axis of the arc is initiated. It is important to recognize that at high currents the diameter of the arc may be greater than the diameter of the nozzle thus leading to the condition known as current choking. When this happens the nozzle is completely blocked and there is no flow of gas. Consequently, the pressure continues to rise due to the continuous change of the volume space V and to the heat energy that is extracted from the arc by the trapped gas.

It is not uncommon to see that when interrupting large currents, specially those corresponding to a three phase fault, the opening speed of the circuit breaker is slowed down considerably due to the thermally generated pressure acting on the underside of the piston assembly.

However, when the currents to be interrupted are low, the diameter of the arc is small and therefore is incapable of blocking the gas flow and as a result a lower pressure is available. For even lower currents, as is the case when switching capacitor banks of just simply a normal load current, it is generally necessary to precompress the gas before the separation of the contacts. This is usually accomplished by increasing the penetration of the arcing contact.

The duration of the compression stroke should always be carefully evaluated to ensure that there is adequate gas flow throughout the range of minimum to maximum arcing time.

In most cases, depending on the breaker design, the minimum arcing time is in the range of 6 to 12 milliseconds. Since the maximum arcing time is approximately equal to the minimum arcing time plus one additional major asymmetrical current loop, which has an approximate duration of 10 milliseconds, then the range of the maximum arcing time is 16 to 22 milliseconds.

What is significant about the arcing time duration is that, since interruption can take place at either of these times, depending only on when a current zero is reached, then what is necessary is that the appropriate pressure be developed at that proper instant where interruption takes place.

It is rather obvious that at the maximum arcing time, the volume has gone through the maximum volume reduction and has had the maximum time expo-

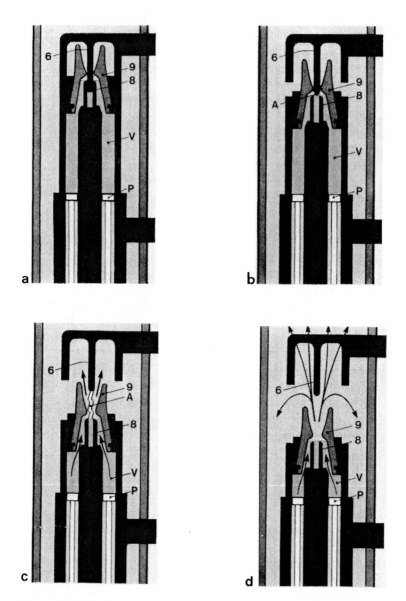

Figure 5.36 Puffer circuit breaker principle (a) breaker closed (b) start of opening, main contacts separate, (c) arcing contacts separating gas flow starts, (d) interruption completed.

Legend: A= Arc, V= Puffer Volume, P= Puffer Piston, 6 & 8 = Arcing Contacts, 9 = Interrupter Nozzle.

sure to the heating action of the arc and thus the gas pressure is expected to be higher. For the minimum arcing time condition, both the compression and the heating of the gas are minimal and therefore the pressure generated is relatively low.

It follows, from the above discussion, that the critical gas flow condition for a puffer interrupter is around the region of the minimum arcing time. However, it also points out that consideration must be given to the opening speed of the breaker in relation to its opening stroke in order to assure that the assumed maximum arcing time is always less than the total travel time of the interrupter.

It was mentioned before that when interrupting large currents in a three phase fault, there is tendency for the breaker to slow down and even to stall somewhere along its opening stroke. This slowing down generally assures that the current zero corresponding to the maximum arcing time is reached before the circuit breaker reaches the end of its opening stroke. However, when interrupting a single phase fault that is not the case. That is, because during a single phase fault the energy input from the fault current is lower which represents a lower generated pressure and so the total force that is opposing the driving mechanism is much lower. Therefore it is quite important to carefully evaluate the single phase operation to assure that there is a sufficient overlap between e the stroke of the puffer (breaker) and the maximum arcing time.

Self Blast Circuit Breakers. Self blast circuit breakers, take advantage of the thermal energy released by the arc to heat the gas and to raise its pressure. In principle the self blast breaker idea is quite similar to the concept of the explosion pot is used by oil circuit breakers. The arc is drawn across a pair of contacts that are located inside of an arcing chamber and the heated high pressure gas is released alongside of the arc after the moving contact is withdrawn from the arcing chamber.

In some designs to enhance the interrupting performance, in the low current range, a puffer assist is added. In other designs a magnetic coil is also included [19]. The object of the coil is to provide a driving force that rotates the arc around the contacts providing additional cooling of the arc as it moves across the gas. In addition to cooling the arc the magnetic coil also helps to decrease the rate of erosion of the arcing contacts and thus it effectively extends the life of the interrupter. In some designs a choice has been made to combine all of these methods for enhancing the interruption. process and in most of the cases this has proven to be a good choice . A cross section of a self blast interrupter pole equipped with a magnetic coils is included in figure 5.37.

5.5.7 Pressure Increase of SF_6 Produced by an Electric Arc

The pressure increase produced by an electric arc burning inside of a small sealed volume (constant volume) filled with SF_6 gas can be calculated with a

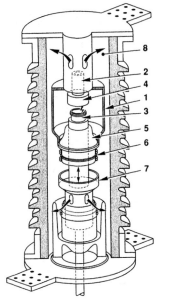

1. Expansion cylinder
2. Fixed arcing contact
3. Moving arcing contact
4. Coil
5. Insulating Spacer
6. Fixed main contact
7. Moving main contact
8. Exhaust volume

Figure 5.37 Outline of a self blast circuit breaker pole.

reasonable degree of accuracy [20] using the curve given in figure 5.38. This curve was obtained by solving the Beattie-Bridgman equation, and by assuming a constant value of 0.8 Joules per gram-degree C for the heat capacity at constant volume (C_v). It is of course this assumption which will introduce some errors because the value of C_v increases with temperature. However the results can be corrected by multiplying the change by the ratio of the assumed C_v to the actual C_v. Values of C_v as a function of temperature are given in figure 5.39.

To calculate the approximate increase in pressure produced by arcing the following procedure may be used.

1. Estimate the arc energy input to the volume. The energy input will be approximately equal to the product of the average arc voltage times the rms. value of the current times the arc time duration.

    For a more accurate calculation the following expression may be used:

$$Q_a = \int_0^t E_a I_m \sin \omega t \; dt$$

where:

Q_a = Arc Energy input in joules
E_a = Arc Voltage
$I_m \sin \omega t$ = Current being interrupted
t = arcing time

2. Find the value of the quotient of the arc energy input, divided by the volume, in cubic centimeters, of the container
3. Find the gas density for a constant volume at normal gas filling conditions using the ideal gas law which says:

$$\rho = \frac{M \times p}{R \times T} \text{ in grams per cubic centimeter}$$

where

M = molecular weight of SF_6 = 146 g
P = Absolute pressure in kilo Pascal
R = Gas constant = 80.5 c.c. -kilo Pascal per mole - °K
T = Temperature degree Kelvin

4. using the just calculated density extract the factor for the pressure rise from figure 5.38 and multiply it by the energy per unit volume obtained in line 1 above.

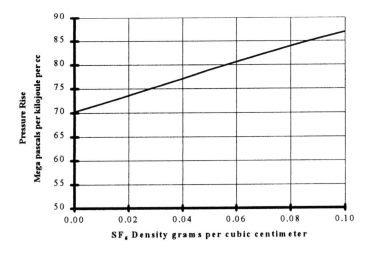

Figure 5.38 Pressure increase for a constant SF_6 volume produced by arcing.

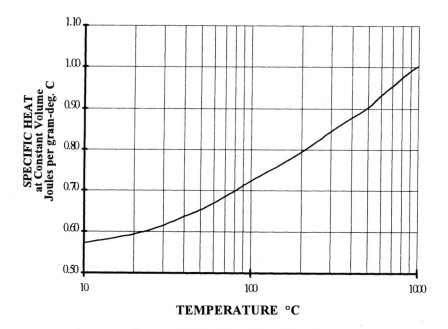

Figure 5.39 Coefficient of heat capacity C_V for SF6 at constant volume.

5.5.8 Parameters Influencing SF₆ Circuit Breaker Performance

Pressure, nozzle diameter, and rate of change of current were the parameters chosen before as the base for evaluating the recovery capabilities of air blast circuit breakers. To facilitate the comparisons between the two technologies the same parameters will now be chosen for SF₆ interrupters and the results are shown graphically in figures 5.40, 5.41 and 5.42. [21]

Once again, the significance of the performance relationships which are shown in the above figures, lies not in their absolute values but in the trends that they predict. In figure 5.40, for example, it is easy to see what is intuitively clear, which is that in order to interrupt larger currents, a larger nozzle diameter is required. It can also be seen the effects of the nozzle diameter and current magnitude; the smaller the current, the lesser the influence of the nozzle diameter. The curve even suggests that there may be a converging point for the nozzle diameters, where at a certain level of smaller currents, the recovery capabilities of the interrupter remains the same regardless of the nozzle size.

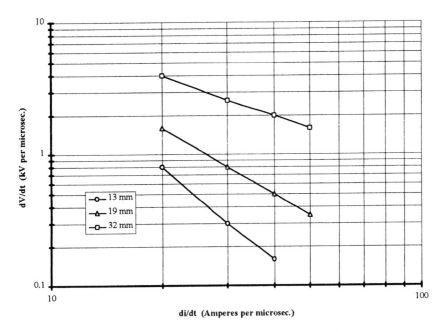

Figure 5.40 Interrupting relationship between current and voltage for various nozzle diameters.

Figure 5.41 shows the dependency of the recovery voltage during the thermal recovery period in relation to the rate of change of current. It is important to note that the slope of each of the lines are remarkably close considering that they represent three independent sources of data extracted from references [7] and [21]. These curves indicate that the rate of recovery voltage in the thermal region is proportional to the maximum rate of change of current (at I = 0) raised to the −2.40 power. The 2.40 exponent compares with the 2.0 exponent obtained with air blast interrupters

In figure 5.42 we find a strong dependency of the recovery on the gas pressure as evidenced by the equation defining the relationship which indicates that the recovery is proportional to the pressure raised to the 1.4 power. In comparison the slope corresponding to the same relationship curves but with air as the interrupting medium is equal to 1.0.

The results observed for the dependency on the rate of change of current and on pressure confirm what we already know; which is, that at the same current magnitude and at the same pressure, SF_6 is a better interrupting medium than air.

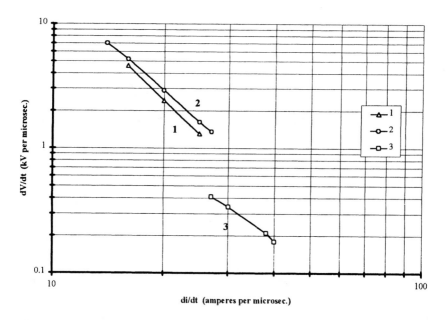

Figure 5.41 Comparison of interrupting capability of SF_6 using data from three independent sources.

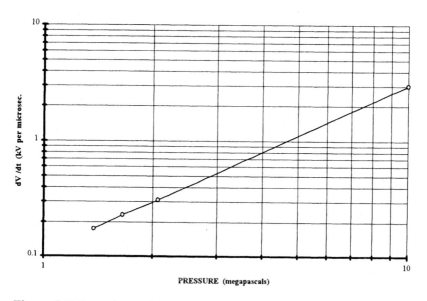

Figure 5.42 Dependency of the recovery voltage upon pressure for SF_6.

5.5.9 SF$_6$-Nitrogen Gas Mixture

Because of the strong dependence of SF$_6$ on pressure it has always been convenient to increase the pressure in order to improve the recovery characteristics of the interrupter. However, as it has been discussed before there are limitations imposed by the operating ambient temperature to avoid liquefaction of the gas.

To overcome the problem the possibility of mixing Nitrogen (N$_2$) with SF$_6$ has been investigated. Although today the issue is only academic when referring to two pressure breakers since they are not manufactured any longer, it has been demonstrated that the performance of a two pressure circuit breaker was improved, as shown in figure 5.43, when at the same total pressure a mixture by pressure of 50 % SF$_6$ and 50 % N$_2$ was used [22], [23].

For single pressure circuit breakers this has not been the case, and this is attributed to the fact that sufficient pressure can not be sustained due to the high flow rates of this lighter gas mixture. In two pressure breakers maintaining the pressure differential high enough is not a problem because the high pressure is maintained in the high pressure reservoir by means of an external compressor.

From the point of view of dielectric withstand no significant difference is found with mixtures containing a high percentage of N$_2$. For example, with a 40% N$_2$ content the dielectric withstand is reduced by only about 10%.

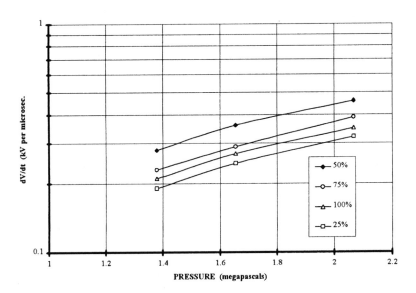

Figure 5.43 Interrupting capability for SF$_6$-N$_2$ mixtures.

5.6 Vacuum Circuit Breakers

Vacuum interrupters take advantage of vacuum because of its exceptional dielectric characteristics and of its diffusion capabilities as an interrupting medium. It should be noted that the remarkable dielectric strength of vacuum is due to the absence of inelastic collisions between the gas molecules which means that there is not an avalanche mechanism to trigger the dielectric breakdown as is the case in gaseous mediums.

The pioneering work on the development of vacuum interrupters was carried out at the California Institute of Technology by R. Sorrensen and H. Mendelhall as reported in their 1926 paper [24].

Despite the early work it was not until the 1950s that the first commercially viable switching devices where introduced by the Jennings Company, and until 1962 when the General Electric Company introduced the first medium voltage power vacuum circuit breaker.

What prevented the earlier introduction of vacuum interrupters where technical difficulties that existed in areas such as the degassing of the contact materials, which is a process that is needed to prevent the deterioration of the initial vacuum due to the release of the gases that are normally trapped within the metals. Another problem was the lack of the proper technologies needed to effectively and reliability weld or braze the external ceramic envelopes to the metallic ends of the interrupters.

In the last 30 years these problems have been solved and that, coupled with the development of highly sensitive instrumentation have substantially increased the reliability for properly sealing the interrupters to prevent vacuum leaks.

In the 1970s there were some attempts made to develop vacuum circuit breakers for applications at voltages greater than 72.5 kV. However these designs were not suitable to compete with SF_6 circuit breakers and vacuum has been relegated primarily to applications in the range of 5 to 38 kV.

In the US vacuum is used most of the time for indoor applications at 5 and 15 kV, and at these or similar voltages it also has the larger share of the market worldwide.

5.6.1 Current Interruption in Vacuum Circuit Breakers

The characteristics of a vacuum, or a low pressure arc, were presented in chapter one. In this section the current interrupting process that takes place in a vacuum interrupter will be described.

Current interruption, like in all circuit breakers, is initiated by the separation of a pair of contacts. At the time of contact part a molten metal bridge appears across the contacts. After the rupture of the bridge a diffuse arc column is formed and the arc is what is called a diffuse mode. This mode is characterized by the existence of a number of fast moving cathode spots, where each spot shares an equal portion of the total current. The current that is carried by the cathode spot depends on the contact material and for copper electrodes a current of about 100 amperes per spot has been observed. The arc will remain in the diffuse column mode until the current exceeds approximately 15 kA. As the magnitude of the current increases a single anode spot appears thus creating a new source of metal vapors which because of the thermal constant of the anode spot continues to produce vapors even after current zero. With the reversal of current, following the passage through zero and because of ion bombardment and a high residual temperature it becomes quite easy to reestablish a cathode spot at the place of the former anode. M. B. Schulman et. al. [25] have reported in the sequence of the arc evolution and have observed that the development is sensitive to the method of initiation.

During normal interruption of an ac current, near current zero the arc column will be diffuse and will rapidly disappear in the absence of current. Since, during interruption and depending in the current magnitude, the arc may undergo the transition from the diffuse mode to the constricted mode and back again to the diffuse mode just prior to current zero it becomes clear that the longer the arc is in the diffuse mode the easier it is to interrupt the current

What it is important to realize from the above is the desirability of minimizing the heating of the contacts and maximizing the time during which the arc remains diffuse during the half current cycle. This objective can be accomplished by designing the contacts in such way that advantage can be taken of the interaction that exists between the current flowing through the arc and the magnetic field produced by the current flowing through the contacts or through a coil that may be assembled as an integral part of the interrupter [26].

Depending in the method used, the magnetic field may act in a transverse or in the axial direction with respect to the arc.

Transverse Field. To create a transverse or perpendicular field different designs of spiral contacts, such as those illustrated in figure 5.44, have been used. In the diffuse mode the cathode spots move freely over the surface of the cathode electrode as if it was a solid disk.. At higher currents and as the arc becomes coalescent the magnetic field produced by the current flowing through the contact spirals forces the arc to move along them [27] as a result of the magnetic forces that are exerted on the arc column as shown in figure 5.45. As the arc rotates its roots also move along reducing the likelihood of forming stationary spots and reducing the localized heating of the electrodes and thus also reducing the emission of metallic vapors. When the end of the contact spirals is reached, the arc roots, due to the magnetic force exerted on the arc column are forced to jump the gap and to continue the rotation along the spirals of the contacts.

Figure 5.44 Two types of spiral contacts used on vacuum interrupters.

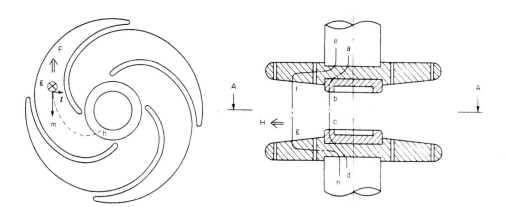

Figure 5.45 Magnetic forces in a transverse magnetic field.

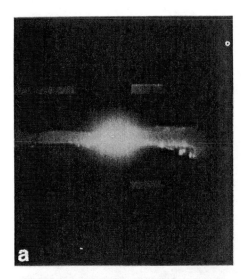

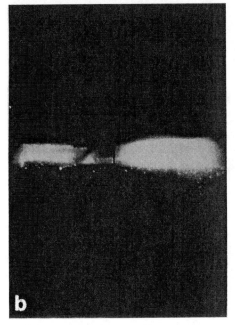

Figure 5.46 Vacuum arc under the influence of a transverse magnetic field: (a) constricted column (18.8 ka peak), (b) arc showing two parallel diffuse columns. (Courtesy of Dr.M. B. Schulman, Cutler-Hammer, Horseheads, NY.)

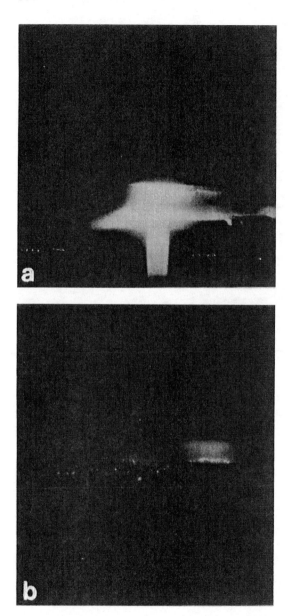

Figure 5.47 vacuum arc under the influence of a transverse magnetic field
(a) jet column with wedge instability (26.8 ka peak), (b) 7 kA diffuse arc following a
current peak of 18.8 kA
(Courtesy of Dr.M. B. Schulman, Cutler-Hammer, Horseheads, NY.)

The effects of the field on the arc are illustrated in figures 5.46 (a), (b) and 5.47 (a), and (b) where the photographs of an arc in the diffuse and constricted modes are shown.

Axial Field. The axial magnetic field decreases the arc voltage and the power input from the arc by applying a magnetic field that effectively confines the arc column as it can be seen in the photograph of a 101 kA peak diffuse arc as shown in figure 5.48.

In the absence of the magnetic field, diffusion causes the arc to expand outwards from the space between the electrodes. However, when the axial magnetic field is present the ion trajectory becomes circumferential and a confining effect is produced. For a reference on the effects of the axial magnetic fields upon the arc column and on the formation of the diffuse arc one can refer to the work published by M. B. Schulman et. al. [28].

Axial magnetic fields can be produced by using either, a coil that is located concentrically outside the envelope of the interrupter and that is energized by the current flowing through the circuit breaker [29], or by using specially designed contacts such as the one suggested by Yanabu et al [30] and which is shown in figure 5.49. Observing at this figure, it can be seen the action of magnetic force on the arc column as the result of the interaction of the magnetic field set up by the current flowing through the arms of the coil electrode and the contact.

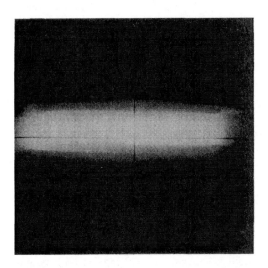

Figure 5.48 High Current (101 kA peak) diffuse arc in an axial magnetic field (Courtesy of Dr. M. B. Schulman, Cutler-Hammer, Horseheads, NY.)

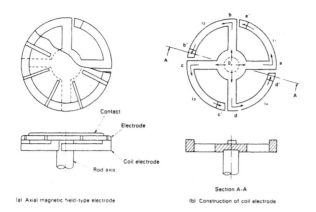

Figure 5.49 Direction of the force on the arc produced by an axial magnetic field.

5.6.2 Vacuum Interrupter Construction

Vacuum interrupters are manufactured by either of two methods. The differences between methods are mainly the procedures used to braze and to evacuate the interrupters.

In one of the methods, which is the one commonly known as the pinch-off method, the interrupters are evacuated individually in a pumping stand after they are completely assembled. An evacuation pipe is located at one end of the interrupter, generally adjacent to the fixed contact and after the required vacuum is obtained the tube is sealed by compression welding.

With the second method the interrupters are concurrently brazed and evacuated in specially designed ovens. The advantage of this method is that evacuation takes place at higher temperatures and therefore there is a greater degree of vacuum purity in the assembly.

The interrupter, as shown in figure 5.50 consists of a ceramic insulating envelope that is sealed at both ends by metallic (stainless steel) plates brazed to the ceramic body so that a thigh vacuum container is created. The operating ambient pressure inside of the evacuated chamber of a vacuum interrupter is generally between 10^{-6} and 10^{-8} torr.

Attached to one of the end plates is the stationary contact, while at the other end the moving contact is attached by means of metallic bellows. The bellows used may be either seamless or welded, however the seamless variety is usually the preferred type.

A metal vapor condensation shield is located surrounding the contacts either inside of the ceramic cylinder, or in series between two sections of the

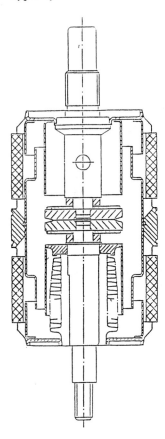

Figure 5.50 Vacuum interrupter construction.

insulating container. The purpose of the shield is to provide a surface where the metal vapor condenses thus protecting the inside walls of the insulating cylinder so that they do not become conductive by virtue of the condensed metal vapor.

A second shield is used to protect the bellows from the condensing vapor to avoid the possibility of mechanical damage. In some designs there is a third shield that is located at the junction of the stationary contact and the end plate of the interrupter. The purpose of this shield is to reduce the dielectric stresses in this region.

5.6.3 Vacuum Interrupter Contact Materials

Seemingly contradicting requirements are imposed upon the possible choices of materials that are to be used for contacts in a vacuum interrupter and therefore the choice of the contact material ends up being a compromise between the requirements of the interrupter and the properties of the materials that are finally chosen [31].

Among the most desirable properties of the contact material are the following:

1. A material that has a vapor pressure that is neither too low nor to high. A low vapor pressure means that the interrupter is more likely to chop the current since there is not enough vapor to maintain the arc at low values of current.

    A high vapor pressure, in the other hand, is not very conducive for interrupting high currents because there would still be a significant amount of vapor remaining at current zero, thus making interruption difficult.

2. A material that has a good electrical conductivity is desired in order to minimize the losses during continuous operation of the interrupter.

3. A high thermal conductivity is also desirable in order to reduce the temperature of the contacts and for obtaining rapid cooling of the electrodes following the interruption of the current.

4. Good dielectric properties are needed to assure rapid recovery capability.

5. High current interruption capabilities.

6. A material that have a low weld strength is needed because contacts in vacuum will invariable weld due to the pre-arcing that occurs when closing or to the localized heating of the micro contact areas when the short circuit current flows through the closed contacts. To facilitate the opening of the contacts easily fractured welds are a basic necessity.

7. Mechanical strength is needed in the material mainly to withstand the impact forces, specially during a closing operation.

8. Materials with low gas content and ease of outgassing are desirable since the contacts must be substantially gas free to avoid the release of any gases from the contacts during interruption and thus to prevent lowering the quality of the vacuum ambient.

9. To prevent the new cathode from becoming a good supplier of electrons a material with low thermionic emission characteristics is desirable.

From the above given list we can appreciate that there are no pure element materials that can meet all of these requirements. Refractory materials such as tungsten offer good dielectric strength, their welds are brittle and thus are easy to break. However, they are good thermionic emitters, they have a low vapor pressure and consequently their chopping current level is high and their interrupting capability is low.

In the other side of the spectrum copper appears to meet most of the requirements. Nevertheless its greatest disadvantage is that due to its ductility it has a tendency to form very strong welds which are the result of diffusion

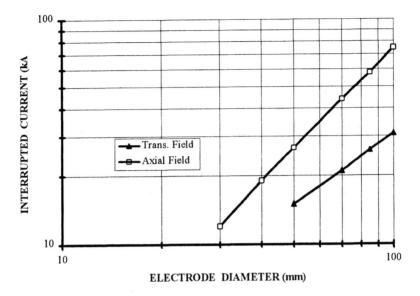

Figure 5.51 Comparison of interruption capability for vacuum interrupters as function of electrode diameter and magnetic field type.

welding. This type of welding occurs, specially inside of a vacuum atmosphere, when two clean surfaces are pushed together and heated..

Since an acceptable compromise material can not be found among the pure elements the attention has been directed to investigate the use of sintered metals or other alloys. A number of binary and ternary alloys have been studied, but from all of those that have been considered two alloys, one a Cu-Bi (copper-bismuth) and the other a Cu-Cr (copper-chrome) alloy have prevailed and today are the most commonly used.

In the Cu-Bi alloy copper is the primary constituent material and the secondary material is bismuth the content of which is generally up to a maximum of 2%. For the Cu-Cr alloy there are different formulations but a typical composition is a 60% Cu to 40% Cr combination.

In general Cu-Bi contacts exhibit a weld strength about 7 times lower than Cu-Cr but they have a higher chopping current level than that of Cu-Cr. The typical chopping level for Cu-Bi contacts is in the range of 3 to 15 amperes with a median value of 7 amperes, while for Cu-Cr is only between 1 to 4 amperes with a median value of 2.7.

Other differences in performance between the materials are the higher rate of erosion that is observed in Cu-Bi contacts and the decrease in dielectric withstand capability that results by the cumulative process of the interrupting duties.

5.6.4 Interrupting Capability of Vacuum Interrupters

From the above discussions it is evident that the interrupting capability of a vacuum interrupter depends more than on anything else on the material and the size of the contacts and in the type of magnetic field produced around the contacts [32]. Larger electrodes in an axial field, as shown in figure 5.51, have demonstrated that they have a better interrupting capability.

Another very important characteristic, related to the interrupting, or recovery capability, of vacuum interrupters is their apparent insensitivity to high rates of recovery voltage [33]. In [7] it is shown that within a frequency range of 60 to 800 Hz., for a given frequency the TRV has only a weak effect on the current magnitude. Furthermore it is widely recognized that the transient voltage recovery capability of vacuum interrupters is inherently superior to that of gas blast interrupters.

REFERENCES

1. ANSI / IEEE C37.100-1981, Definition for power Switchgear.
2. Practical Application of Arc Physics in Circuit Breakers. Survey of Calculation Methods and Application Guide, Electra (France) No. 118: 65-79, May 1988.
3. ANSI C37.06-1979 Preferred Ratings and Related Capabilities for ac High Voltage Circuit Breakers Rated on a Symmetrical Current Basis.
4. International Electrotechnical Commission (IEC), International Standard IEC 56.
5. G. Frind, Time Constant of Flat Arcs Cooled by Thermal Conduction, IEEE Transactions on Power Apparatus and Systems, Vol. 84, No.12: 1125-1131, Dec. 1965.
6. Emil Alm, Acta Polytechnica, L47 (1949) Electrical Engineering Series Vol. 2, No. 6, UDC 621.316.5.064.2, Sweden Royal Academy of Engineering Sciences.
7. D. Benenson, G. Frind, R. E. Kissinger, H. T. Nagamatsu, H. O. Noeske, R. E. Sheer Jr., Fundamental Investigations of Arc Interruption in Gas Flows, Electric Power Research Institute, EPRI-EC-1455, 1980.
8. Current Interruption in High Voltage Networks, (ed. K. Ragaller) Plenum Press, New York, 1978.
9. Roy Wilkins, E. A. Cretin, High Voltage Oil Circuit Breakers, Mac. Graw Hill, 1930.
10. F. Kesselring, Theoretische Grundlagen zur Berechnung der Schaltgeräte, Walter de Gruyter & Co., Berlin, 1968.

11. H. Moissan, P. C. LeBeau, Royal Acad. Sci. 130: 984-988, 1900.
12. H. J. Lingal, A. P. Strom, T. E. Browne, Jr., An Investigation of the Arc Quenching Behavior of Sulfurhexaflouride, AIEE Trans. 72, Pt. III: 242-246, 1953.
13. Allied Signal, Accudri SF$_6$, Technical Bulletin 97-0103.4M.595M, 1995
14. Elizabeth Cook, Lifetime Commitments: Why Policy-Makers Can't Afford to Overlook Fully Flourinated Compounds, Issues & Ideas, World Resources Institute, Feb. 1995.
15. EPA, Electrical Transmission and Distribution Systems, Sulfurhexaflouride and Atmospheric Effects of Green House Gas Emissions, Conference Final Report, Aug. 9-10, 1995.
16. Malkom K. W. Ko, Nien Dak Sze, Wei-Chyung Wang, George Shia, Aaron Goldman, Frank J. Murcray, David J. Murcray, Curtiss P. Rinsland, Atmospheric Sulfurhexaflouride: Sources, Sinks and Greenhouse Warming, Journal of Geophysical Research, Vol. 98, No. D6: 10499-10507, June 20, 1993.
17. G. Mauthe, K. Petterson, R. Probst, H. Bräutigam, D. Köning, L. Niemayer, B. M. Pryor, CIGRE 23.10 Report.
18. Isunero Ushio, Isad Shimura, Shotairo Tominaga, Practical Problems of SF$_6$ Gas Circuit Breakers, PA&S Vol. PAS-90, No.5: 2166-2174, Sept/Oct. 1971.
19. G. Bernard, A. Girard, P. Malkin, An SF$_6$ Auto Expansion Breaker: The Correlation Between Magnetic Arc Control and Critical Current, IEEE Trans. on Power Del., Vol. 5, No. 1: 196-201, Jan. 1990.
20. R. Garzon, Increase of Pressure in a Vessel Produced by an Electric Arc in SF$_6$, ITE Internal Engineering Report 3032-3.002-E7, 1973.
21. R. D. Garzon, Rate of Change of Voltage and Current as Function of Pressure and Nozzle Area in Breakers Using SF$_6$ in the Gas and Liquid Phases, IEEE Transactions of Power Apparatus and Systems, Vol PAS-95, No. 5: 1681-1688, Sep./Oct. 1976.
22. R. D. Garzon, The Effects of SF$_6$-N$_2$ Mixture Upon the Recovery Voltage Capability of a Synchronous Interrupter, IEEE Transactions on Power Apparatus and Systems, Vol PAS-95, No. 1: 140-144, Jan./Feb. 1976.
23. Wang Erzhi, Lin Xin, Xu Jianyuan, Investigation of the Properties of SF$_6$/N$_2$ Mixture as an Arc Quenching Medium in Circuit Breakers, Proc. of the 10th. Int. Conf. on Gas Disch. and their Appl., Vol. 1: 98-101, Swansea, UK, Sept. 1992.
24. R. W. Sorensen, H. E. Mendenhall, Vacuum Switching Experiments at the California Institute of Technology, Transactions AIEE 45,: 1102-1105, 1926.
25. M. Bruce Schulman, Paul Slade, Sequential modes of Drawn Vacuum Arcs Between Butt Contacts for Currents in the 1 kA to 16 kA Range,

194   *Chapter 5*

IEEE Trans. on Components, Packaging, and Manufacturing Technology-Part A, Vol. 18N0. 2: 417-422, June 1995.

26. R. Gebel, D. Falkenberg, Behavior of Switching Arc in Vacuum Interrupters Radial Field and Axial Field Contacts, ITG- Fachber (West Germany), Vol. 108: 253-259, 1989.

27. M. Bruce Schulman, Separation of Spiral Contacts and the Motion of Vacuum Arcs at High AC Currents, IEEE Trans. on Plasma Scien, Vol. 21, No. 5:484-488, Oct. 1993.

28. M. B. Schulman, Paul G. Slade, J. V. R. Heberlein, Effect of an Axial Magnetic Field Upon the Development of the Vacuum Arc Between Opening Electric Contacts, IEEE Trans. on Components, Hybrids, and Manufacturing Technology, Vol. 16, No. 2, 180-189, March 1993.

29. H. Schellekens, K. Lenstra, J. Hilderink, J. terHennere, J. Kamans, Axial Magnetic Field Type Vacuum Circuit Breakers Based on Exterior Coils and Horse Shoes, Proc. XII th, INT. Symp. on Diel. Disch. and Elec. Insulation, Cat. No. 86CH2194-9: 241-244, Shoresh, Israel, Sept. 1986.

30. S. Yanabu, E Kaneko, H. Okumura, and T. Aiyoshi, Novel Electrode Structure of Vacuum Interrupter and its Practical Application, IEEE Trans. Power App & Syst. Vol. PAS 100: 1966-1974, March-April 1981.

31. P. Slade, Contact Materials for Vacuum Interrupters, IEEE Trans. On Parts, Hybrids, and Packaging, Vol. PHP-10, No. 1, March 1974.

32. Toshiba, Technical Bulletin, KSI-E1052-2, 1983-6.

33. R. K. Smith, Test Show Ability of Vacuum Circuit Breaker to Interrupt Fast Transient Recovery Voltage Rates of Rise of Transformer Secondary Faults, IEEE Trans. on Power Del., Vol. 10, No, 1: 266-273, Jan. 1995.

34. Allan Greenwood, Vacuum Switchgear, IEE Power Series 18, Short Run Press Ltd. Exeter England, 1994.

6

MECHANICAL DESIGN OF CIRCUIT BREAKERS

6.0 Introduction

The two most basic functions of a circuit breaker are to open and close their contacts on command. This at first sight implies a rather simple and trivial task; however, many of the characteristics involved in the process of opening, closing and maintaining the contacts closed can be quite demanding.

It is interesting to note that according to a CIGRE report [1] more than 90 % of circuit breaker failures are attributed to mechanical causes. These findings confirm the fact that circuit breakers are primarily mechanical devices that are called upon to perform an electric function.

The majority of the time circuit breakers remain closed and simply act as electrical conductors, but in many occasions they do indeed perform their intended protective functions and when this happens, from the combined electrical and mechanical point of view, undoubtedly the contact structure is probably one of the most essential and critical components. A second and equally important component is the operating mechanism employed to produce the motion of the contacts.

These two components are closely linked to each other and in more ways than one they determine the success or failure of an interrupting device. Given the importance of these two components this chapter will be dedicated to the discussion of subjects relating to these components.

The most commonly used designs of operating mechanisms will be described in general terms, concentrating primarily in describing the operational sequences, rather than dealing with specific details of how to design a particular type of a mechanism.

The subject of electrical contacts will be treated in more detail so that a better understanding is gained on this area of design which tends to re-occur frequently, not only when dealing with the development of new circuit breakers but when evaluating circuit breaker performance or special applications.

6.1 Contact Theory

Circuit breaker contacts must first be able to carry their assigned continuous current rating, without overheating, or deteriorating and must do so within reasonable limits of power consumption.

In addition, during short circuit conditions, they must be able to carry large currents for some specified periods of time, and again they must do so without deteriorating or arcing.

To meet these requirements it is indispensable that among other things the resistance of the contacts be kept as low as possible, that the contact area be maximized, that the materials are properly selected for the application at hand, that proper contact force be applied, that the optimum number of contacts be selected, that the contact cross section and the contact mass are properly sized, and that the minimum operating speed, both during closing and during opening, are sufficient to limit erosion of the contacts.

6.1.1 Contact Resistance

The resistance of a clean, ideal contact, where any influence due to oxide films is neglected and where it is assumed that a perfect point of contact is made at a spot of radius (*r*), is given by the following equation:

$$R = \frac{\rho}{2r}$$

where:

R = Contact resistance

ρ = Resistivity of contact material

r = Radius of contact spot

However, in actual practice, this is not the case and the real area of contact is never as simple as it has been assumed above. It should be recognized that no matter how carefully the contact surfaces are prepared the microscopic interface between two separable contacts invariably will be a highly rough surface, having a physical contact area that is limited to only a few extremely small spots. Furthermore, whenever two surfaces touch they will do so at two micro points where even under the lightest contact pressure, due to their small size of these points, will cause them to undergo a plastic deformation that consequently changes the characteristics of the original contact point.

It is clear then that contact force and actual contact area are two important parameters that greatly influence the value of contact resistance. Another variable that also must be taken into consideration is the effects of thin films, mainly oxides, that are deposited along the contact surfaces.

6.1.1.1 Contact Force

When a force is applied across the mating surfaces of a contact the small microscopic points where the surfaces are actually touching are plastically deformed and as a result of this deformation additional points of contact are established. The increase in the number of contact points serves to effectively decrease the value of the contact resistance.

The contact force F, exerted by a pair of mating contacts can be approximated by the following equation [1]

$$F = k \times H \times A_r$$

where:

   H = Material hardness

   A_r = Contact area

   k = Constant between 0.1 and 0.3

The constant of proportionality k is first needed to account for the surface finishes of the contacts and secondly because, in reality, the hardness is not constant since there are highly localized stresses at the micro points of contact.

6.1.1.2 Contact Area

Even though it can not be determined very accurately, the knowledge of the approximate areas of contact is essential for the proper understanding and design of electrical contacts. It is found that contact resistance is a function of the density of the points of contact as well as of the total area of true contact within the envelope of the two engaging full contact surfaces. In a well-distributed area the current will diffuse to fill all the available conducting zone, but in practical contacts this area is greatly limited because it is not possible to have such degree of precision on the alignment, nor is it possible to attain and maintain such high degree of smoothness.

In the earlier discussion, dealing with contact pressure, it was implied that the contact area is determined solely by the material hardness and by the force that is pushing the contacts together.

Since the original simplified equation for the contact resistance was given in terms of resistivity of the material and the radius of the contact point it is then possible to substitute the term for the contact radius with the expression for the contact force, noting that:

$$A_r = \pi a^2$$

When this is done and the term R_F representing the film resistance is added the final expression for the total contact resistance R_T then becomes:

$$R_T = \frac{\rho}{2} \sqrt{\frac{\pi k H}{F}} + R_F$$

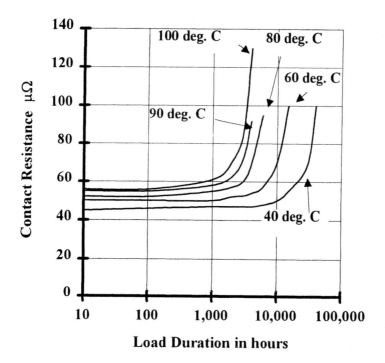

Figure 6.1 Resistance run-away condition as a function of time of carrying load currents for copper contacts that are immersed in oil a various temperatures (ref. 4 © IEEE)

6.1.2 Insulating Film Coatings on Contacts

A pure metal to metal contact surface can only be achieved in a vacuum, or in an inert gas atmosphere. In air the contact surfaces can oxidize and they become coated with a thin oxide film. According to Holm [2] a layer of 3 to 30 microns is formed on copper in a few seconds and almost instantly on aluminum, while it takes about two days to form on a silver or silver plated surface.

If the formed oxides are insulating, as is the case of CuO in copper contacts, then due to their build-up the contact spots will gradually reduce in size thus decreasing the contact area and increasing the contact resistance. This process, as observed by Williamson [3] and by Lemelson [4] and as shown in figure 6.1 is not very noticeable in its early stages but in its later stages it will suddenly get into a run away condition.

The formation of a sulfide coat on a silver contact surface also produces an increase in contact resistance. This situation can develop on SF_6 circuit breakers after the contacts have been subjected to arcing and when there is no

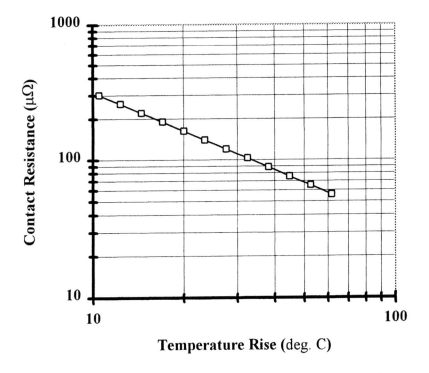

Figure 6.2 Change in contact resistance as a function of temperature rise at the contacts. Silver plated contacts exposed to sulfides resulting as by-products of arcing in SF_6.

scraping, or wiping motion between the contacts. However in references [5] and [6] it is pointed out that the sulfide film on a silver surface is easily removed by slight friction and that it may even become decomposed by heat. The later has been demonstrated experimentally and the results are shown in figure 6.2. It can be seen in this figure that the resistance is reduced as a function of the temperature rise which in turn was reached by passing 600 amperes through the interrupter.

6.1.3 Contact Fretting

Fretting is described as an accelerated form of oxidation that takes place across the contact surfaces and that is caused by any continuous cyclical motion of the contacts. Initially the junction points of the contact spots will seize and will eventually shear, however, this shear action does not increase the contact resistance because the particles are of pure metal. As the cycle repeats and the metal fatigue progresses then the metal layers are softened and separate allowing the oxide layer to grow until contact is lost.

The increase of contact resistance under these conditions has been observed as being a strong function of current, contact force and plating material.

To avoid this problem it is important then, when designing a contact interface to consider using silver plating specially on aluminum bars.

6.1.4 Temperature at the Point of Contact

Because of the analogy that exists between the electric and the thermal fields and if the assumption is made that there is no heat loss by radiation in the close proximity of a contact the following relationship, relating the voltage drop measured across the contact and its temperature, can be established.

$$\theta = \frac{V_C^{\,2}}{8\lambda\rho}$$

where:

θ = Temperature
λ = Thermal conductivity of contact material
ρ = Electric resistivity of contact material
V_C = Voltage drop across contact

This equation however is valid only within a certain limited range of temperatures. At higher temperatures the materials, at the contact interfaces, will begin to soften and thus can undergo a plastic deformation. At even higher temperatures the melting point of the material will be reached. In table 6.1 below the softening and the melting temperatures together with their corresponding voltage drop are tabulated.

The significance of the above is that now we can determine for a specific material the maximum currents at which either softening or melting of the contacts would occur and consequently the proper design to avoid the melting and welding of the contacts can be made.

The equations for the maximum softening and melting currents are:

$$\text{Softening current } I_s = \frac{2V_s}{\rho}\sqrt{\frac{F}{\pi k H}} \quad \text{and}$$

$$\text{Melting current } I_m = \frac{2V_m}{\rho}\sqrt{\frac{F}{\pi k H}}$$

TABLE 6.1
Softening and Melting Temperatures
for Contact Materials

Contact	Softening		Melting	
Material	Temperature K.	Voltage Drop mV	Temperature K	Voltage Drop mV
Au	373	80	1336	450
Ag	423	90	1233	350
Al	423	10	931	300
Zn	443	10	692	170
Cu	463	120	1356	430
Ni	793	220	1728	650
Pt	813	250	2046	700
Mo	1172	340	2883	960
W	1273	400	3653	1000

6.1.5 Short Time Heating of Copper

The maximum softening and melting currents as defined by the above equations are applicable to the point of contact and are useful primarily for determining the contact pressure needs. However when dealing with the condition where the contacts are required to carry a large current for a short period of time it is useful to define a relationship between time, current and temperature for different materials.

Below is given a general derivation for a general expression that can be used for determining the temperature rise in a contact.

First, it will be assumed that for very short times all the heat produced by the current is stored in the contact and is therefore effective in producing a rise in temperature.

Then the heat generated by a the current (i) flowing into a contact of (R) Ohms during a dt interval is:

$$Q = R\,i^2\,dt \quad \text{(Joules) and}$$

The heat required to raise the contact temperature by d(degrees C is:

$$Q = S\,V\,d\theta \quad \text{(Joules)}$$

Since it was assumed that there is no heat dissipation then it is possible to write

$$R\,i^2\,dt = S\,V\,d\theta$$

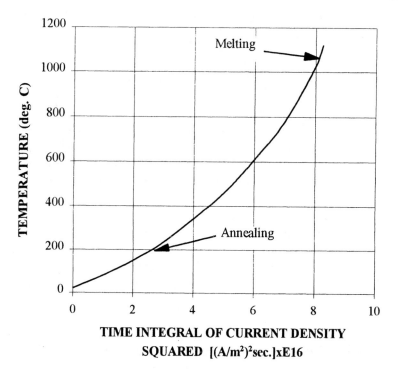

TIME INTEGRAL OF CURRENT DENSITY
SQUARED [(A/m²)²sec.]xE16

Figure 6.3 Short time heating of copper as function of the time integral of the current

squared ($\theta = \int\limits_{0}^{t} J^2 dt$)

where:

i = current in amperes
t = time in seconds
S = specific heat of material in joules per m³ per °C
θ= temperature in °C
R = contact resistance in Ohms
V = contact volume in m³

It can be shown that substituting the resistance of the contact with the specific resistivity of the material as a function of a standard ambient temperature which is assumed to be 20 °C, and by integrating the function the following equation is obtained:

$$\frac{1}{A^2}\int\limits_{0}^{t} i^2 dt = 11.55 \times 10^{16} \log_{10} \frac{234+\theta}{254}$$

by substituting the current density it is possible to re-write the equation as:

$$\theta = f\left(\int J^2 dt\right)$$

The graph shown in figure 6.3 represents a general curve for the approximate heating of a copper contact of uniform cross section, with a current that starts to flow when the initial contact temperature is 20 °C. Similar curves can be generated for other contact materials by using the proper constant for the new material under consideration.

Since it is known the softening, or melting temperatures for a given material then it is possible to determine, from the graph the integral of the current density and furthermore since the current is constant we have:

$$\int J^2 dt = J^2 t$$

and therefore

$$t = \frac{\int J^2 dt}{J^2}$$

6.1.6 Electromagnetic Forces on Contacts

As it has been described before, we know that there is a current constriction at the point of contact. We also know that this constriction is responsible for the contact resistance and consequently for the heat being generated at the contacts, but in addition to this the current constriction is also the source of electromagnetic forces acting upon the contact structures. In figure 6.4 the current path of the current at the mating of the contact surface is shown and as it can be seen from the figure, as the current constricts into a transfer point, there is a component of the current that flows in opposite directions and thus the net result is a repelling force trying to force apart the contacts.

According to Holm [2] the repulsion force is given by

$$F_R = 10^{-7} I^2 \ln\frac{B}{a} \quad \text{Newton}$$

where:

B = the contact area
a = actual area of contact point

It is difficult however to properly use this formula because of the difficulty in defining the actual contact area. For practical purposes some expressions that yield adequate results have been proposed. One such expression is given by Greenwood [7] as:

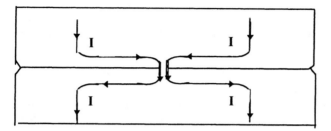

Figure 6.4 Current constriction at the actual point of contact.

$$F_R = 0.112 \times \left(\frac{I}{n}\right)^2 \quad \text{lb. per finger}$$

Another practical relationship which was obtained experimentally with a 3 inch Cu-Cr butt type contact in vacuum is shown in figure 6.5 and is given by the expression:

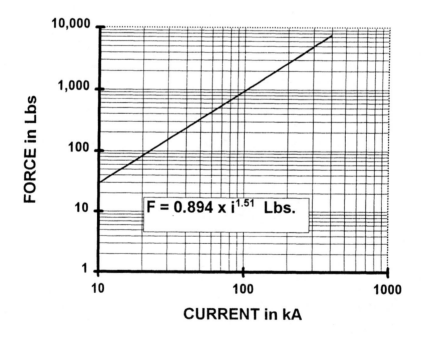

Figure 6.5 Measured blow-out force for 3 in. Cu-Cr butt contacts in vacuum.

$$F_R = 0.885 \times \left(\frac{I}{n}\right)^{1.51} \text{ lb. per finger}$$

where:

I = the peak current in kA
n = the number of contacts

The differences on the force requirements obtained depending on which expression is used serve to point out the uncertainty of the variables involved and the probabilistic nature of the forces. In general the application of higher forces would yield a higher confidence level and a higher probability of withstanding the repulsion forces.

6.1.6.1 Force on Butt Contacts

The repulsion forces acting on butt contacts, such as those used in vacuum interrupters, can be calculated using any of the expressions given in the previous paragraph, considering the number of contacts n as being equal to 1. The magnitude of the repulsion forces must be counteracted by the operating mechanism and therefore a proper determination of the force magnitude is essential for the design and application of the operating mechanism.

6.1.6.2 Force on Circular Cluster Contact

When a contact structure, such as the one shown in figure 6.6 (a), and (b) is used, it can be shown that in addition to the repelling force there is another force. This additional force is due to the attraction between a set of two opposite fingers where current is flowing in the same direction on each contact (fig. 6.6 (a)). The attraction or blow-in force for a circular contact configuration can be calculated using the following equation:

$$F_A = 0.102(n-1)\left(\frac{I}{n}\right)^2\left(\frac{l}{d}\right) \text{ Newton}$$

6.1.6.3 Force on a Non-circular Non-symmetrical Cluster Contact

The following method can be used to calculate the blow-in force for each of any of the parallel contacts in the arrangement illustrated in figure 6.6 (c). The forces are calculated assuming that the current divides equally among each contact and although this is not totally accurate the results are close enough to provide an indication of the suitability of the design for an specific application.

$$F_{1-5} = 0.102\frac{l}{a}\left(\frac{I}{n}\right)^2\cos\alpha \quad \text{Newton}$$

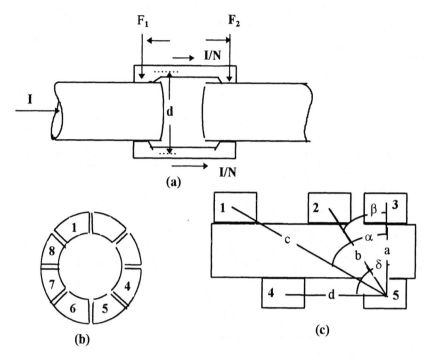

Figure 6.6 Diagram of blow-in forces relationships for circular and non-symmetrical contacts.

$$F_{2-5} = 0.102 \frac{l}{b} \left(\frac{I}{n} \right)^2 \cos \beta \qquad \text{Newton}$$

$$F_{3-5} = 0.102 \frac{l}{c} \left(\frac{I}{n} \right)^2 \cos \gamma \qquad \text{Newton}$$

$$F_{4-5} = 0.102 \frac{l}{d} \left(\frac{I}{n} \right)^2 \cos \delta \qquad \text{Newton}$$

6.1.6.4 Total Force On Contacts

The total force acting on a contact therefore becomes:

$$F_T = F_S + F_A - F_R$$

where:

F_T = Total force per contact segment
F_S = Contact spring force

F_A = Blow-in, or attraction force per contact segment
F_R = Blow-out, or repulsion force per contact segment

In a properly designed contact it would be expected that $F_A \geq F_R$

6.1.7 Contact Erosion

Contact erosion is the unavoidable consequence of current interruption and is caused primarily by the vaporization of the cathode and the anode electrodes. According to W. Wilson [9] the heating leading to the vaporization at the electrodes is the result of the accompanying voltage drops. He derived an equation for the vaporization rate in cubic centimeters of contact material lost per kA of current. In figure 6.7 the results of tests reported in reference [9] have been reproduced. The graph presents the rate of erosion in air for various contact materials. The data is plotted in what is called "their order of excellence," that is from best to worst, best being the material exhibiting the least amount of erosion.

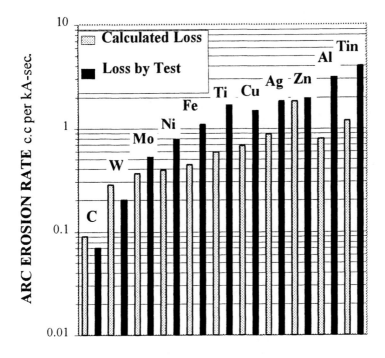

Figure 6.7 Contact material erosion due to arcing (data from ref. 8).

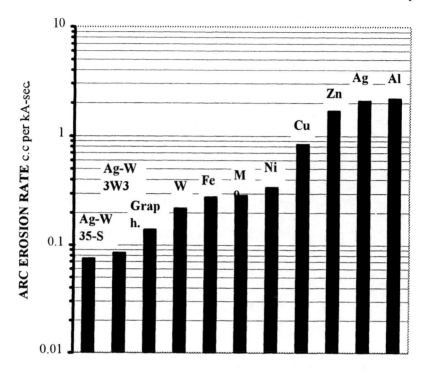

Figure 6.8 Measured materials rate of erosion due to arcing in SF$_6$.

In figure 6.8 the rate of contact erosion in SF$_6$ based on a set of personal unpublished data is plotted. It is interesting to note that there is a reasonable degree of agreement between these values and those given by Wilson. As calculated and as "vaporized loss by test."

It seems therefore that the following equation can be used to obtain reasonable estimates for the rate of erosion of the contacts.

$$R = \frac{1000\left(E_C + E_A\right)}{\rho JH}$$

where:

      R = erosion rate, cc per kA – sec.
      E_C = cathode drop, volts
      E_A = anode drop, volts
      J = heat equivalent, 4.18 joules per calorie
      H = heat of vaporization, calories per gram
      ρ = density of contact material, grams per cc

6.2 Mechanical Operating Characteristics

Opening and closing velocities, as well as stroke, or travel distance are the most important operating characteristics of a circuit breaker. They are dictated primarily by the requirements imposed by the contacts. Opening and closing velocities are important to the contacts in order to avoid contact erosion as well as contact welding. Circuit breaker stroke is primarily related to the ability of the circuit breaker to withstand the required operating dielectric stresses which are directly related to the total contact gap.

6.2.1 Circuit Breaker Opening Requirements

Two basic requirements for the total opening operation of a circuit breaker are the opening speed and the total travel distance of the contacts. The opening speed requirements are dictated by the need to assure that the parting of the contacts is done as rapidly as possible for two reasons; first, to limit contact erosion and second by the need to control the total fault duration which is dictated by the system coordination requirements. The total travel distance is not necessarily the distance needed to interrupt the current but rather the gap space needed to withstand the normal dielectric stresses and the lighting impulse waves that may appear across the contacts of a breaker that is connected to a system while in the open position.

The need for carrying the continuous current and for withstanding a period of arcing, makes it necessary to use two sets of contacts in parallel. One, the primary contact, which is always made of a high conductivity material such as copper and the other, the arcing contact, made of synthetic arc resistance materials like copper or silver tungsten or molybdenum which have a much lower conductivity than those used for the primary contacts.

By having a parallel set of contacts when one of these contacts opens, due to the differences in the resistance and the inductance of the electrical paths, there is a finite time that is required to attain total current commutation, that is, from the primary or main contact branch to that of the arcing contact.

The significance of the commutation time can be appreciated when one considers that in the worst case commutation may not take place until the next current zero is reached and that during this time the arc is eroding the copper main contacts. Since arc erosion of the contacts not only limits the life of the contacts but it can also lead to dielectric failures by creating an ionized conducting path between the contacts and thus limiting the interrupting capability of the circuit breaker. It is also important to realize that commutation must be completed before the arcing contacts separate, otherwise, the arc is likely to remain at the main contacts.

The commutation time can be calculated by solving the electrical equivalent circuit for the contact arrangement as given in figure 6.9.

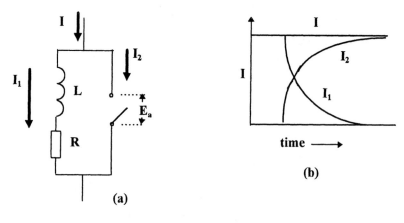

Figure 6.9 Equivalent circuit for determining the required arc commutation time between main and arcing contacts.

It can be shown that:

$$I = I_1 + I_2 = I_m\sin(\omega t + \phi) \quad \text{and}$$

$$I_1 = I - \frac{E_a}{R}\left(1 - e^{-\alpha t}\right.$$

where:

$$\alpha = \frac{R}{L}$$

I = system current
I_1 = main contact current
I_2 = arcing contact current
R = arcing contact resistance
L = loop inductance of arcing contact path

Commutation is completed at t = t_1 when $I_2 = I$

The commutation time is then obtained by solving the above equation for the time t_1:

$$t_1 = \frac{L}{R}\ln\left(\frac{1}{1 - \dfrac{IR}{E_a}}\right)$$

From the above it is observed that to have a successful commutation of current it requires that $E_a > IR$.

While the opening operation continues and as the contact gap increases a critical contact position is reached. This new position represents the minimum contact opening where interruption may be accomplished at the next current zero. The remainder of the travel is needed only for dielectric and deceleration purposes.

Under no-load conditions and when measured over the majority of the travel distance, the opening and closing velocity of a circuit breakers is constant, as it is shown in 6.10 (a) and (b) where actual measurements of the circuit breaker speeds are shown.

Although different type circuit breakers have different speed and travel requirements, the characteristic shape of the opening travel curve are very much similar in all cases. The average speed is usually calculated by measuring the slope of the travel curve over the region defined by the point of initial contact part to a point representing approximately three fourths of the total travel distance.

The opening speed for vacuum circuit breakers is generally specified in the range of 1 to 2 meters per second, for SF_6 circuit breakers the range is in the order of 3 to 6 meters per second.

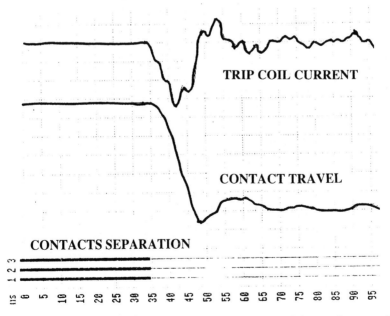

Figure 6.10 (a) Typical distance vs. time measurement of the opening operation of a circuit breaker.

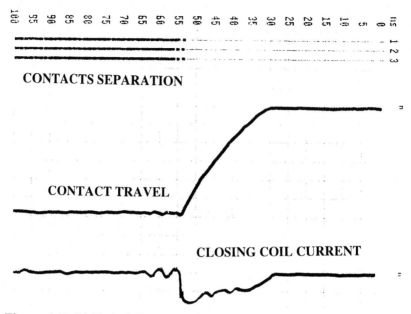

CONTACTS SEPARATION

CONTACT TRAVEL

CLOSING COIL CURRENT

Figure 6.10 (b) Typical distance vs. time measurement of the closing operation of a circuit breaker.

For vacuum circuit breakers there is another, often overlooked, requisite for the initial opening velocity of the contacts which corresponds with the assumed point of the critical gap distance. Considering that the typical recovery voltage of vacuum is about 20 to 30 kV per millimeter then for those applications at 25 or 38 kV it is extremely important that at an assumed 2 milliseconds minimum arcing time the actual gap should be about 3 millimeters which translates into a velocity of 3 meters per second and even though this is a relatively modest velocity in comparison to other circuit breakers the fact that butt contacts are used in vacuum interrupters means that we need a higher initial acceleration since there is no contact motion prior to the actual beginning of the contact separation. In lieu of contact wipe or contact penetration, the overtravel, or contact spring wipe, that is provided for compensation of contact erosion and to create a hammer or impact blow needed to break the contact weld can be utilized as a convenient source of kinetic energy to deliver a high rate of initial acceleration to the contacts. However because of the mechanics at the time of impact it is usually recommended that the mass of the mechanism moving parts be equal to at least twice the mass of the moving contact.

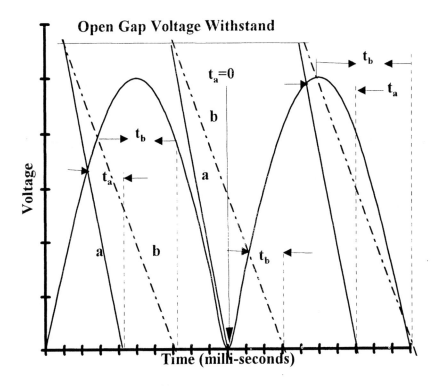

Figure 6.11 Prestrike during closing relationship between closing velocity and system voltage.

6.2.2 Closing Speed Requirements

During a closing operation and as the contacts approach each other, a point is reached where the gap equals the minimum flashover distance and therefore an electric arc is initiated. As the distance between the contacts continues to diminish the arc gradually shortens until finally the contacts engage and the arc disappears. Therefore as we have seen not only when opening but also during closing an arc can appear across a pair of contacts.

Depending upon the voltage, the interrupting medium and the design of each particular circuit breaker, the contact flashover characteristics vary very widely. However, let us assume two characteristic slopes as shown in figure 6.11(slopes *a* and *b*). Now, when these slopes are superimposed on the plot of the absolute values of a sinusoidal wave that represent the system voltage, it shows, that depending on the instantaneous relation between the contact gap and the system voltage the arc is initiated at the point of intersection of the two curves.

The elapsed time between the flashover point and the time where the contacts engage represents the total arcing time which is shown as t_a and t_b corresponding to slopes a and b respectively. It also can be seen in this figure that the arcing time decreases as the slope of the flashover characteristics increases which suggests what should be obvious that increasing the closing velocity decreases the arc duration.

Increasing the closing velocity not only decreases the arcing time but it also decreases the magnitude of the current at the instant of contact engagement. Assuming that the electromagnetic repulsion forces are a function of the peak current squared then as show by P. Barkan [9] the work done by the mechanism against the electromagnetic repulsion is:

$$\varepsilon = \int_0^T I_m^2 \left(\sin^2 \omega t \right) V dt$$

The dimensionless solution of this equation is given as:

$$\frac{\varepsilon}{2kI_m^2 S} = 1 - \left(\frac{V}{2\omega S} \right) \sin\left(\frac{2\omega s}{V} \right)$$

where:

ε = work done against electromagnetic force

k (0.112 = constant of proportionality for electromagnetic force

I_m = peak current at contact touch

S = contact travel

V = contact velocity

ω = 377 radians for 60 Hz currents

The benefits of a high closing speed are then a reduction in the mechanism energy requirements and a reduction of the contact erosion.

6.3 Operating Mechanisms

The primary function of a circuit breaker mechanism is to provide the means for opening and closing the contacts. At first this seems to be a rather simple and straightforward requirement but, when one considers that most circuit breakers once they have been placed in service will remain in the closed position for long periods of time and yet in the few occasions when they are called upon to open or close they must do so reliably, without any added delays, or sluggishness; then, it is realized that the demands on the mechanisms are not as simple as first thought.

It is important then to pay special attention to such things as the type of grease used, the maximum stresses at the latch points and bearings, the stiffness of the whole system and most of all to the energy output of the mechanism.

Just as there are different types of circuit breakers, so are there different types of operating mechanisms, but, what is common to all is that they store potential energy in some elastic medium which is charged from a low power source over a longer period of time.

From the energy storage point of view the mechanisms that are used in today's circuit breakers fall in the spring, pneumatic or hydraulic categories, and from the mechanical operation point of view they either of the cam or of the four bar linkage type.

6.3.1 Cam versus Linkage

Cams are generally used in conjunction with spring stored energy mechanisms and these cam-spring driven mechanisms are mostly used to operate medium voltage vacuum interrupters.

Cam drives are flexible, in the sense that they can be tailored to provide a wide variety of motions, they are small and compact. However, the cam is subjected to very high stresses at the points of contact, and furthermore the cam follower must be properly constrained, so that it faithfully follows the cam's contour, either by a spring which raises the stress level on the cam, or by a grooved slot where the backlash may cause problems at high speeds. For those interested in further information for the design of a spring operated cam-follower system Barkan's paper [10] is highly recommended.

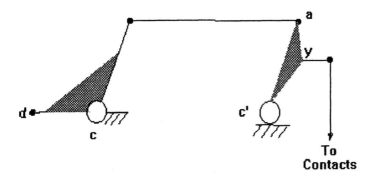

Figure 6.12 Schematic diagram of a four bar linkage arrangement.

Linkages, in most cases, have some decided advantages over cams; for one they are more forgiving in terms of fabrication accuracy because small variations on their lengths do not significantly influence their motion. In general the analysis of a four bar linkage system is a relatively simpler task and it is not that difficult to determine the force at any point in the system.

Since the mechanism must deliver as much work as it receives, then when friction is neglected, the force at any point multiplied by the velocity in the same direction of the force at that point must be equal to the force at some other point times the velocity at that same point, or in other words the forces are inversely proportional to their velocities.

Four bar linkages such as the one shown schematically in figure 6.12 are practically synonymous with pneumatic and hydraulic energy storage mechanisms.

6.3.2 Weld Break and Contact Bounce

Contact bounce and contact welding are two conditions which are usually uniquely associated with vacuum circuit breakers.

We already know that vacuum contacts weld upon closing and that therefore it is necessary to break the weld before the contacts can be opened. Breaking the weld is accomplished by an impact, or hammer blow force which is applied, preferably directly to the contacts. To provide this force it is a common practice to make use of the kinetic energy that is acquired by the opening linkages as they travel over the compressed length of the wipe springs.

To assure that the proper impact force will be available, the velocity at the time of contact separation can be used as a guideline.

$$< \frac{M_1 \dfrac{dx_1}{dt} + M_2 \dfrac{dx_2}{dt}}{M_1 + M_2}$$

where:

M_1 = Contact's mass

M_2 = Mechanism's mass (moving parts)

x_1, x_2 = displacement

Contact bounce can have serious effects on the voltage transients generated during closing and it may cause damage to equipment connected to the circuit breaker, although protective measures can be taken to reduce the magnitude of the transients it is also advisable to design the circuit breaker so that the bounce is at least reduced, if not eliminated.

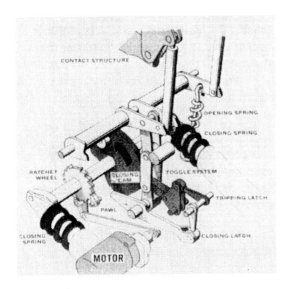

Figure 6.13 Simplified drawing of a spring stored energy mechanism.

Suppression of bounce requires that a close interaction be established between the energy dissipation at the contact interface and the energy storage of the supporting structure of the interrupter and its contacts. If means are provided for the rapid dissipation of the stored energy in the supporting structure the effect of the bounce can be minimized.

6.3.3 Spring Mechanisms

A simplified drawing depicting a typical spring type operating mechanism is show in figure 6.13. This type of mechanism is commonly found in some medium voltage outdoor and in practically all of the medium voltage indoor type circuit breakers. However, it is not uncommon to find these mechanisms also on outdoor circuit breakers up to 38 kV and in a few cases they even have been used on 72.5 kV rated circuit breakers.

As its name suggests the energy of this mechanism is stored on the closing springs. The stored energy is available for closing the circuit breaker upon command following the release of a closing latch.

The spring mechanism, in its simplest form, consists of a charging motor and a charging ratchet, a closing cam, closing springs, opening springs and a toggle linkage. The charging motor and ratchet assembly provide automatic recharging of the closing springs immediately following the closing contact sequence.

The charged springs are held in position by the closing latch which prevents the closing cam from rotating. To release the spring energy either an electrically operated solenoid closing coil, or a manual closing lever is operated. Following the activation of the closing solenoid a secondary closing latch is released while the primary latch rotates downward due to the force being exerted by the charged closing springs and thus the rotation of the closing cam which is connected to the operating rods is allowed. As the cam rotates it straightens the toggle linkage (refer to figure 6.13), which in turn rotates the main operating shaft thus driving the contacts that are connected to the shaft by means of insulating rods.

The straightening of the toggle links loads the trip latch as they go over-center. The trip latch then holds the circuit breaker in the closed position. In addition to closing the contacts the closing springs supply enough energy to charge the opening springs.

Opening of the contacts can be initiated either electrically or manually; however, the manual operation is generally provided only for maintenance purposes. When the tripping command is given the trip latch is released freeing the trip roller carrier. The force produced by the over center toggle linkage rotates the trip roller carrier forward, and as the first toggle link rotates about its pivot it releases the support that provided to the second and third links. The opening springs which are connected to the main operating shaft provide the necessary energy to open the contacts of the circuit breaker.

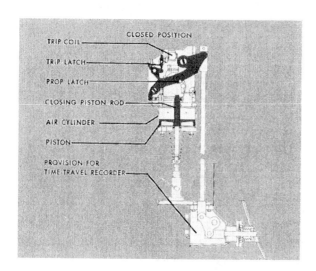

Figure 6.14 Illustration of a pneumatic mechanism.

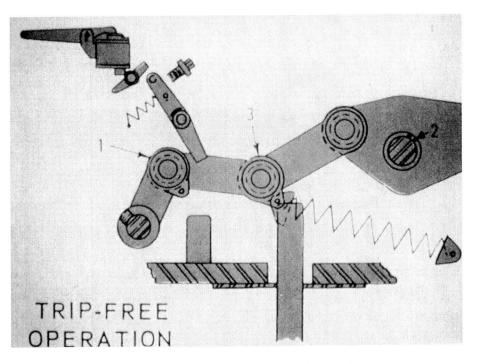

Figure 6.15 Toggle linkages arrangement to satisfy trip free requirements.

6.3.3 Pneumatic Mechanisms

Pneumatic mechanisms are a logical choice for air blast circuit breakers and that is so because pressurized air is already used for insulating and interrupting; however, pneumatic mechanisms are not limited to air blast breakers, they also have been used to operate oil and SF_6 circuit breakers.

Those mechanisms, which are used with air blast circuit breakers, usually open and close pneumatically and in some cases there is only a pneumatic rather than a solid link connection between the mechanism and the contacts.

Other pneumatic mechanisms, such as the one that is illustrated in figure 6.14, use an air piston to drive the closing linkage and to charge a set of opening springs. This mechanism, which has been used in connection with oil and SF6 circuit breakers, has a separate air reservoir where sufficient air is stored at high pressure for at least 5 operations without need of recharging in between operations.

Figure 6.16 Photograph of a hydraulic mechanism used by ABB T&D Co.

To close the circuit breaker, high pressure air is applied to the underside of the piston by opening a three way valve, the piston moves upwards transmitting the closing force through a toggle arrangement, figure 6.15, that is used to provide the trip free capability, to the linkage which is connected to the contacts by means of an insulating push-rod. In addition to closing the contacts the mechanism charges a set of opening springs and once the contacts are closed a trip latch is engaged to hold the breaker in the closed position.

Opening is achieved by energizing a trip solenoid which in turn releases the trip latch thus allowing the discharge of the opening springs which forces the contacts to the open position..

Another variation of a pneumatic mechanism is one where the pneumatic force is used to do both, the closing and the opening operation. The direction being controlled by the activation of either of the independent opening or closing three way valves.

6.3.4 Hydraulic Mechanisms

Hydraulic mechanisms are in reality only a variation of the pneumatic operator, the energy, in most cases, is stored in a nitrogen gas accumulator and the incompressible hydraulic fluid becomes a fluid operating link that is interposed

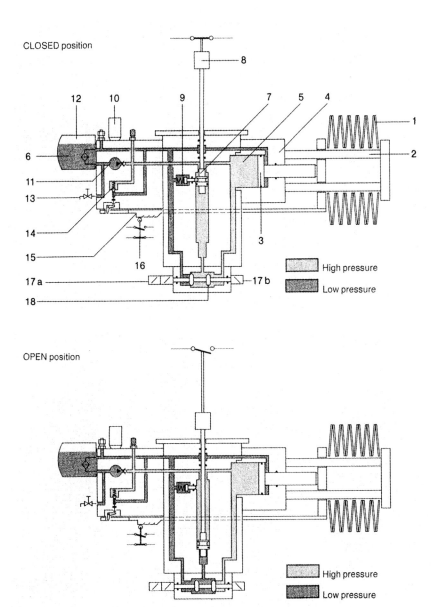

Figure 6.17 Functional operating diagram of the mechanism shown in figure 6.16.

between the accumulator and a linkage system that is no different than that used in conjunction with pneumatic mechanisms.

In a variation of the energy's storage method, the nitrogen accumulator is replaced by a disk spring assembly which acts as a mechanical accumulator. A mechanism of this type is shown in figure 6.16. It offers significant advantages; it is smaller, there is no chance for gas leaks from the accumulator, and the effects of the ambient temperature upon the stored energy are eliminated.

The operation of this mechanism can be described as follows; (note that the numerical references to components correspond to those shown in figure 6.17).

A supply of hydraulic oil is filtered and stored in a low pressure reservoir (12), from where it is compressed by the oil pump (11). The high pressure oil is then stored in reservoir (5). The piston (3) which is located inside of the high pressure storage is connected to the spring column (1). The springs are supported by the tie bolts (2). A control link (15) checks on the charge of the spring column and activates the auxiliary switch contacts (16) that control the pump's motor (10) as required to maintain the appropriate pressure.

With the circuit breaker in the closed position the operating piston (7), which is connected to the conventional circuit breaker linkages (8), has high pressure applied to both of its faces. To open the breaker the opening solenoid (17a) is energized causing the changeover valve to switch positions and connect the underside of the operating piston to low pressure (6), thus causing the piston to move to the open position. Closing as the reverse of the opening and is initiated by energizing the closing solenoid (17b) and by admitting high pressure to the underside of the operating piston. Item (4) is an storage cylinder, (9) is a mechanical interlock (13) is an oil drain valve and (14) is a pressure release valve.

REFERENCES

1. CIGRE SC13, High Voltage Circuit Breaker Reliability Data For Use In System Reliability Studies, CIGRE Publication, Paris France 1991.
2. R. Holm Electrical Contacts, Almquist & Wiksells Akademiska Handböcker, Stockholm, Sweden, 1946.
3. J. B. P. Williamson, Deterioration process in electrical connectors, Proc. 4th. Int. Conf. Electr. Contact Phenomena, Swansea, Wales, 1968.
4. K. Lemelson, About the Failure of Closed Heavy Current Contact Pieces in Insulating Oil at High Temperatures, IEEE Trans. PHP, Vol. 9, March 1973.
5. G. Windred, Electrical Contacts, McMillan and Co. Ltd., London, 1940.
6. S. C. Killian, A New Outdoor Air Switch And A New Concept Of Contact Performance, AIEE Trans. Vol. 67: 1382-1389, 1948.

7. A. Greenwood, Vacuum Switchgear, IEE Power Series 18, London, UK, 1994.

8. W. R. Wilson, High-Current Arc Erosion of Electrical Contact Materials, AIEE Trans. Part III, PAS Vol. 74: 657-664, August 1955.

9. T. H. Lee, Physics and Engineering of High Power Devices, The Massachusetts Institute of Technology: 485-489, 1975.

10. P. Barkan, R. V. McGarrity, A Spring-Actuated Cam Follower System; Design Theory and Experimental Results, Trans. of ASME, Journal of Engineering for Industry, Vol. 87, Series B, No. 3: 279-286, 1965.

7

A COMPARISON OF HIGH VOLTAGE CIRCUIT BREAKER STANDARDS

7.0 Introduction

The preferred ratings, the performance parameters and the testing requirements of circuit breakers are well established subjects that are covered by a number of applicable standards. These standards, both nationally and internationally, have evolved over a long period of time and this evolution process undoubtedly will continue because of the necessity to reflect the changing needs of the industry.

Future changes in the standards can be expected, not only because of changes in the circuit breaker technology but, due to better, faster, more advanced and more accurate relaying and instrumentation packages and also because of the much broader use of digital equipment that is now replacing old electromagnetic-analog devices.

In this chapter we will review the most influential current international standards, their historical background, their organization and procedures, as well as their specific requirements and the reasons for these choices. We will discuss these current standards covering ratings and performance parameters; comparing their differences and providing clarification as to the intent and the origin of the ratings.

The section dealing with power testing of circuit breakers deserves special attention and therefore the next chapter will be dedicated completely to this subject.

7.1 Recognized Standards Organizations

The two most recognized and influential circuit breaker standards in the world are supported by the American National Standards Institute (ANSI) and the International Electrotechnical Commission (IEC).

ANSI is the choice document in the US, and in places around the world where the US has had a strong influence in the development of their electric industry. IEC standards are invoked by all other countries outside of the sphere of US influence, and today the majority of circuit breakers that are being sold worldwide are being built to meet the IEC standards

The format of the two standards is significantly different, and to some extent there are noticeable differences in the technical requirements, but these differences are not so radical to the point of being non-reconcilable. These differences have more to do with local established practices than with fundamental theory and if there is something to be said about experience, then both standards have demonstrated to be more than adequate for covering the needs of the industry. The most significant differences will be discussed in the appropriate sections as we review the generic requirements.

A current issue that commands a great deal of attention is the issue of harmonization. This has become increasingly more important not only because of agreements reached by the World Trade Organization and the globalization of trade, but, also because today all of the basic development of high voltage SF_6 circuit breakers is being done overseas and all of the US suppliers of this type of equipment are owned by European or Japanese multinational corporations.

7.1.1 ANSI/IEEE/NEMA

ANSI is a member's federation that acts as a coordinating body for all volunteer standards writing organizations in the US with the aim of developing global standards that reflect US interests. ANSI provides the criteria and the process for approving a consensus.

High voltage circuit breakers and switchgear standards are, for the most part, developed through the combined and separate effort of IEEE and NEMA. Both of these groups are co-secretariats of the Accredited Standards Committee C37 (ASC C37) which serves as the administrator, or the clearing house through which the proposed standards are submitted to ANSI for publication as an American National Standard. It is the Accredited Standards Committee who assigns the industry's well known and readily identifiable series C37 numbers to the ANSI/ IEEE/NEMA circuit breakers standard.

The membership of the technical committees, subcommittees and working groups of the IEEE consists of voluntary individuals that represent users, manufacturers and interested groups, in a reasonably well balanced proportion. The membership of NEMA consists solely of equipment manufacturers.

To help visualize the inter-relationships that exists between these organizations and the flow of the standard documents, a basic organizational diagram is included in figure 7.1.

The first report on standardization rules in the electrical field can be traced to 1899 when, it was prepared by the American Institute of Electrical Engineers, (AIEE), now IEEE. In the years that followed, other organizations, such as the Electric Power Club, which in 1926 merged with the Associated Manufacturers of Electrical Supplies to become what is now NEMA, became interested in the process of standardization. In today's scheme, NEMA is responsible for the development of those standards that are related to Ratings, Construction and Conformance Testing. IEEE responsibility is to develop all other technical standards related to Requirements, Performance, and Testing including design, production, and field testing.

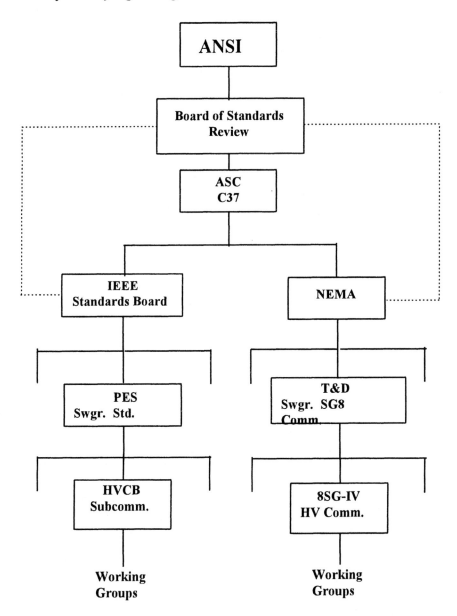

Figure 7.1 Organizational Chart of ANSI/IEEE and NEMA Groups Dealing with High Voltage Circuit Breaker Standards.

The main ANSI high voltage circuit breaker standards are (NEMA) C37.06 AC High-Voltage Circuit Breakers Rated on a Symmetrical Current Basis- Preferred Ratings and Related Required Capabilities, ANSI/IEEE C37.04 IEEE Standard Rating Structure for AC High-Voltage Circuit Breakers Rated on a Symmetrical Current Basis, and ANSI/IEEE C37.09 IEEE Standard Test Procedure for AC High-Voltage Circuit Breakers Rated on a Symmetrical Current Basis.

7.1.2 IEC

IEC, as an organization, dates back to the early 1900's. Its membership consists of national delegates from member countries which are organized in the form of national committees. Presently there are forty-nine (49) national committees and one (1) associate member country. Each national committee, or member nation have a single vote This is one important point to emphasize, because it shows IEC as being a truly international organization where the decisions are approved at the country levels and each country member has only one vote. This represents the major philosophical difference between IEC and ANSI, since ANSI is only a national organization where each individual has one vote.

The United States National Committee (USNC) to the IEC serves as the ANSI sponsored US delegation. Within this delegation there are a number of technical advisory groups (TAGs) that are called upon to support the individuals who act in the role of technical adviser (TA). It should be noted that both IEEE and NEMA are heavily involved and have a strong influence within the USNC delegations because, most of its members are also members of either or both IEEE and NEMA.

IEC high voltage circuit breaker standards are prepared by the Technical Committee 17 (TC 17). More specifically, the scope of TC 17 is to prepare international standards regarding specifications for circuit breakers, switches, contactors, starters, disconnectors, busbar and any switchgear assemblies. Subcommittee 17 A (SC 17A) is responsible for the current applicable standards IEC 56 (1987) and IEC 694 (1980) Common clauses for high-voltage switchgear and controlgear standards.

The organizational relationships within IEC are illustrated in the accompanying figure 7.2.

7.2 Circuit Breaker Ratings

The ratings of a circuit breaker, given by the applicable standards, are considered to be the minimum designated limits of performance that are expected to be met by the device. These limits are applicable within specified operating conditions. In addition of including the fundamental voltage and current parameters, they list other additional requirements which, are derived from the

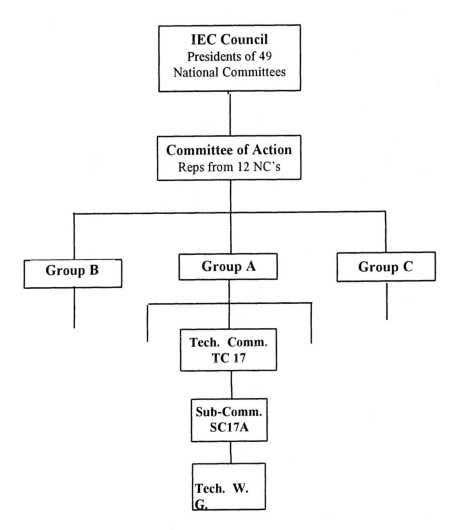

Figure 7.2 Organizational Chart of IEC Groups Dealing with High Voltage Circuit Breaker Standards.

above listed basic parameters, and which in the ANSI documents are identified as related required capabilities.

ANSI C37.06 contains a number of tables where a list of Preferred Ratings is included. These ratings are just that, "preferred", because they are the ones more commonly specified by users and are simply those which have been selected by NEMA strictly for convenience, and in order to have what it could be said to be an off the shelf product. The fact that there is a listing of preferred ratings does not exclude the possibility of offering other specific ratings as required, provided that all the technical performance requirements and the related capabilities specified in the appropriate standard (C37.04) are met.

7.2.1 Normal Operating Conditions

ANSI considers as normal or usual operating condition ambient temperatures which do not exceed plus 40 °C and which are not below minus 30°C, and altitudes which do not exceed 3300 ft or 1000 m. ANSI does not differentiate service conditions between indoor or outdoor applications.

IEC does differentiate for indoor, or outdoor applications. It specifies the altitude limit at 1000 m and the maximum ambient temperature as plus 40 °C for both applications; however it additionally specifies that the average of the maximum temperature over a 24 hour period does not exceed 35 °C.

For the lower temperature limits there are two options given for each application. For indoor, there is a minus 5 °C limit for a class "minus 5 indoor" and a minus 25 °C for a class "minus 25 indoor." For outdoor applications there is a class "minus 25" and a class "minus 40." Limits for icing and wind velocity are also recognized by IEC but are not by ANSI.

7.2.2 Rated Power Frequency

This seemingly simple rating that relates only to the frequency of the ac power system has a significant influence when related to other circuit breaker ratings. The power frequency rating is a significant factor during current interruption, because for many types of circuit breakers, the rate of change of current at the zero crossing is a more meaningful parameter than the given rms. or peak current values. In all cases, when evaluating the interrupting performance of a circuit breaker, it should be remembered that, a 60 Hz current is generally more difficult to interrupt than a 50 Hz current of the same magnitude

7.2 Voltage Related Ratings

7.2.1 Maximum Operating Voltage

Whether called rated maximum operating voltage (ANSI) or rated voltage (IEC), this rating sets the upper limit of the system voltage for which the breaker application is intended.

The selection of the maximum operating voltage magnitudes is based primarily on current local practices. Originally, ANSI arrived at the maximum operating voltage rating by increasing the normal operating voltage by approximately 5 % for all breakers below 362 kV and by 10 % for all breakers

Table 7.1
ANSI and IEC Rated Operating Voltages
(Voltages shown in kV)

For voltages 72.5 kV and below									
IEC	3.6	7.2	12	17.5	24	36	52	72.5	
ANSI	4.76	8.25	15	15.5	25.8	38	48.3	72.5	
For voltages above 72.5 kV									
IEC	100	123	145	170	245	300	362	420	525	765
ANSI		121	145	169	242		362		550	800

above 362 kV. However, the practice of using the nominal operating voltage has been discontinued, primarily, because the use of only the maximum rated voltage has become common practice by other related standards, [1], [2], for apparatus that are used in conjunction with circuit breakers..

As it can be seen below, in Table 7.1, the preferred values that are specified in each of the two standards are relatively close to each other. The voltage ratings of 100, 300 and 420 kV are offered in the IEC due to the practices in Europe, while these ratings are not common in the US and therefore are not offered by ANSI. Furthermore, it is expected that in the next revision of the preferred ratings for outdoor oilless breakers, which are listed in ANSI C37.06, the values for 121, 169 and 242 kV will be changed to harmonize with the IEC values.

7.2.2 Rated Voltage Range Factor K

The rated voltage range factor, as defined by ANSI, is the ratio of the rated maximum voltage, to the lower limit of the range of operating voltage, in which the required symmetrical and asymmetrical interrupting capabilities vary in inverse proportion to the operating voltage. This is a rating that is unique to ANSI, and it is a carry over from earlier standards that were based on older technologies such as oil, and air magnetic circuit breakers; where, as we know, a reduced voltage results in and increase in the current interrupting capability. With modern technologies, namely vacuum and SF_6 this is no longer applicable. This fact was recognized for outdoor oils circuit breakers, and the rated range factor for this type of circuit breakers was eliminated more than twenty years ago.

Another reason that has been given for the adoption of the K factor rating is the convenience of grouping a certain range of voltages under a common denominator, which in this case was chosen to be a constant MVA. Where, MVA is equal to: $\sqrt{3}$ × rated maximum voltage × rated symmetrical current.

Currently there is a motion that is working its way through the standards committees seeking to delete this requirement as it applies to indoor circuit breakers.

7.2.3 Rated Dielectric Strength

The minimum rated dielectric strength capability of a circuit breaker is specified, by the standards, in the form of a series of tests that must be performed, and which simulate power frequency overvoltages, surge voltages caused by lighting, and overvoltages resulting from switching operations.

7.2.3.1 Low frequency dielectric

The withstand capability is one of the earliest established rating parameters of a circuit breaker. AIEE standards, as early as 1919, specified a one minute, 60 Hz., dry tests and selected a value of 2.25 times rated voltage plus 2000 volts as the basis of this rating. In addition to the dry test, a 10 second wet test, which still is required, was included. The voltage magnitude chosen for this test was 2 times rated voltage plus 1000 volts.

In the IEC standards, the low frequency withstand voltages, for circuit breakers rated 72.5 kV and below, match the one minute dry test values specified by ANSI; However, at voltages above 72.5 kV the IEC requirements are significantly lower than the ANSI ones. For circuit breakers rated 242 kV and below IEC does not have a 10 second wet test requirement as ANSI does.

Since, in general any power frequency. overvoltages that may occurs in an electric system are much lower than the low frequency, (60 Hz) withstand values that are required by the standards, we can only conclude that these higher margins where adopted in lieu of switching surge tests which, at the time were not specified and consequently, were not performed.

7.2.3.2 Lighting impulse withstand

These requirements are imposed in recognition to the fact that overvoltages produced in an electric system by lighting strokes are one of the primary causes for system outages and for dielectric failures of the equipment. The magnitude and the waveform of the voltage surge, at some point on a line, depends on the insulation level of the line and on the distance between the point of origin of the stroke and the point on the line which is under consideration.

This suggests that it is not only difficult to establish a definite upper limit for these overvoltages, but that it would be impractical to expect that high voltage equipment, including circuit breakers, should be designed so that they are capable of withstanding the upper limits of the overvoltages produced by lighting strokes. Therefore, and in contrast to the 60 Hz requirements; where the margins are very conservative and are well above the normal frequency overvoltages that may be expected, the specified impulse levels are lower than the levels that can be expected in the electrical system in the event of a lighting stroke, whether it is a direct stroke to the station, or the most likely event, which is a stroke to the transmission line that is feeding the station.

The objective of specifying an impulse withstand level, even though is lower that what can be seen by the system, is to define the upper capability limit for a circuit breaker and to define the level of voltage coordination that must be provided.

The Basic Impulse Level, (BIL), that is specified, in reality only reflects the insulation coordination practices used in the design of electric systems, and which are influenced primarily, by the insulation limits and the protection requirements of power transformers and other apparatus in the system. Economic considerations also play an important role in the selection because, in most cases the circuit breaker must rely only on the protection that is offered by surge protection that is located at a remote location from the circuit breaker and close to the transformers and because as is common practice surge arresters at the line terminals of the circuit breaker are generally omitted. In reference [3] C. Wagner et. al., have described a study they conducted to determine the insulation level to be recommended for a 500 kV circuit breaker. The study, as reported, was done by equating the savings obtained by increasing the insulation levels, against the additional cost of installing additional surge arresters, if the insulation levels were to be lowered. The findings, in the particular case that was studied, suggested that a 1300 kV level was needed for the 500 kV transformers in the installation, however, a level of 1550 kV was chosen as the most economic solution for the associated circuit breakers and disconnect switches. Other studies have shown that in some cases, for a similar installation, an 1800 kV BIL would be needed. The findings of the two evaluations then, further suggested, that two values may be required, but having two designs to meet the different levels is not the optimum solution from a manufacturing point of view and consequently only the higher value was adopted as the standard rated value.

ANSI specifies only one BIL value for each circuit breaker rating with the exception of breakers rated 25.8 and 38 kV where two BIL values are given. The lower value is intended for applications on a grounded wye distribution system equipped with surge arresters. IEC, in contrast, specifies two BIL ratings for all voltage classes, except for 52 and 72.5 kV where only one value is given, and for 245 kV where three values are specified.

The comparative values for circuit breakers rated 72.5 kV and above are given in Table 7.2. From this table it can be seen that up to the 169/170 kV rating the ANSI values are directly comparable to the higher value given for each voltage class by IEC, at 245 kV, for all practical purposes, the 900 kV

Table 7.2
BIL Comparison
ANSI and IEC

ANSI					IEC	
Rated Voltage kV	p.u. of rated voltage	BIL	2 μsec. Chopped Wave	3 μsec. Chopped Wave	BIL	Rated Voltage kV
72.5	4.8	350	452	402	325	72.5
121	4.55	550	710	632	550 450	123
145	4.5	650	838	748	650 550	145
169	4.45	750	968	862	750 650	170
242	3.7	900	1160	1040	1050 950 850	245
362	3.58	1300	1680	1500	1175 1050	362
550	3.26	1800	2320	2070	1550 1425	550
800	2.56	2050	2640		2100 1800	800

required by ANSI should be adequate for the two lower values of IEC. As is seen at 362 and 550 kV the ANSI values are significantly higher than the IEC values.

The preceding comparison shows the variability of the requirements and reaffirms the fact that the values, as chosen, are generally adequate whenever proper coordination procedures are followed a fact that is corroborated by the reliable operating history of the equipment. It can also be observed that up to 169 kV ratings the per unit ratio between the impulse voltage and the maximum voltage of the breaker is basically a constant of approximately 4.5 p.u. but as the rated voltage of the breaker increases the BIL level is decreased. C. Wagner [4] attributes this decrease to the grounding practices, since at these voltage levels all systems are effectively grounded, and also to the types of surge arresters used.

7.2.3.3 Chopped wave withstand

This dielectric requirement is specified only in ANSI and it has been a part of these standards since 1960. This requirement was added in recognition of the fact, that the voltage at the terminals of the surge arrester has a characteristically flat top appearance, but at some distance from the arrester, the voltage is somewhat higher. This characteristic had already been taken into account by transformer standards where a 3 µsec chopped wave requirement was specified.

An additional reason for establishing the chopped wave requirement was to eliminate, primarily for economic reasons, the need for surge arresters at the line side of the breaker and thus, to allow the use of rod gaps and to rely on the usual arresters which are used at the transformer terminals. The 3 µsec rating is given as 1.15 times the corresponding BIL. This value happens to be the same as that of the transformers and it assumes that the separation distance between the arrester and the circuit breaker terminals is similar as that between the transformer and the arrester. The 2 µsec peak wave is given as 1.29 times the corresponding BIL of the breaker. The higher voltage is intended to account for the additional separation from the breaker terminals to the arresters in comparison to the transformer arrester combination.

Basic Lightning Impulse Tests. The tests are made under dry conditions using both, a positive and a negative impulse wave. The standard lighting impulse wave is defined as a 1.2 × 50 microseconds wave. The waveform and the points used for defining the wave [5], are shown in figure 7.3 (a). The 1.2 µsec value represents the front time of the wave and is defined as 1.67 times the time interval t_f that encompasses the 30 and 90 % points of the voltage magnitude, when these two points are joined by a straight line. The 50 µsec represents the tail of the wave and is defined as the point where the voltage has declined to

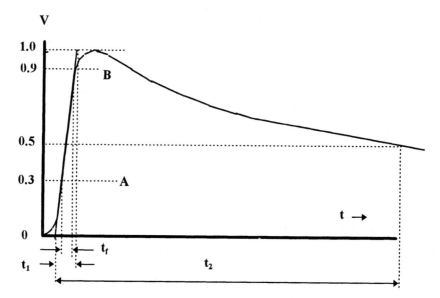

Figure 7.3 (a) Standard 1.2 × 50 Impulse Wave.

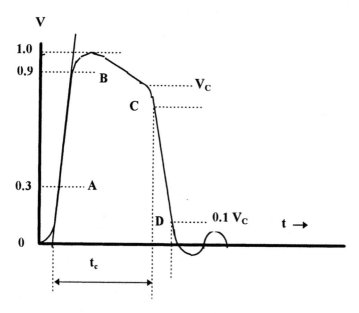

Figure 7.3 (b) Standard Wave Form for Chopped Wave Tests.

half its value. The time is measured from a point t = 0 which is defined by the intercept of the straight line between the 30 and 90 % values and the horizontal axis that represents time.

In figure 7.3 (b) a chopped wave is illustrated, the front of the wave is defined in the same manner as above but, the time shown as t_c, which represents the chopping time, is defined as the time from the wave origin to the point of the chopping initiation.

Until recently, ANSI has been specifying a 3 × 3 test method, which meant that the impulse wave is applied three consecutive times to each test point, if a flashover occurred during the three initial tests, then three additional tests had to be performed where no flashovers were allowed.

The test method has now been changed and the new requirement defines a 3 x 9 test matrix. With this procedure, a group of three tests are performed, if one flashover occurs then, a second set of nine tests, where no flashovers are allowed, must be performed. The reason for the change in the test procedure was based on the desire to increase the confidence level on the withstand capability of the design.

IEC specifies a 2 × 15 method which requires that a group of fifteen consecutive tests be made and where only two flashovers, across self restoring insulation, are allowed during the series. However, IEC has given the option for employing the 3 × 9 method, which can be accepted subjected to the agreement between the user and the manufacturer.

Confidence Level Comparison. To evaluate which test yielded a higher confidence level, the approach to the statistical comparison of the methods that was used is given below.

The Binomial Distribution is used to evaluate the required confidence levels, this distribution considers a number of independent experiments, n, each with a given probability of success, p, and probability of failure, $(1 - p)$. If x is the number of successes and $(n - x)$ is the number of failures, then the probability of a random variable, X, being greater or equal to some x_0 is:

$$\Pr(X(x_0)) = \sum_{x=x_0}^{n} \frac{n!}{(n-x)!\,x!} p^x (1-p)^{n-x}$$

and the confidence level is given by

$$C = 1 - \Pr(X(x_0))$$

For the 2 x 15 method the confidence level is calculated as follows:

$n = 15$ tests

$x \geq x_0 = 13$ successes minimum

$p = 90\%$ probability of success on any test

$$\Pr \geq X\,(13) = \sum_{x=13}^{15} \frac{15!}{(15-x)!\,x!}(0.9)^x(1-0.9)^{15-x}$$

$\Pr \geq X\,(13) = 0.816$ and

$C = 1 - \Pr(X \geq 13) = 1 - 0.816 = 0.184$

This indicates that there is an 18.4 % confidence level that the breaker has a 90 % probability of withstanding the lighting impulse.

Now for the 3×9 test we must first find the confidence level if 3 out of 3 tests are successful

$n = 3$ tests

$x \geq x_0 = 3$ successes minimum

$p = 90\%$ probability of success on any test

$$\Pr \geq X\,(3) = \sum_{x=3}^{3} \frac{3!}{(3-x)!\,x!}(0.9)^x(1-0.9)^{3-x} = 0.729$$

and $C = 1 - \Pr \geq X\,(3) = 1 - 0.729 = 0.271$

So there is a 27.1 % confidence level that the breaker has a 90 % probability of withstanding the lighting impulse.

Next, the confidence level is calculated if one failure occurs in the first 3 tests and none in the remaining 9

$n = 12$ tests

$x \geq x_0 = 11$ successes minimum

$p = 90\%$ probability of success on any test

$$\Pr \geq X\,(11) = \sum_{x=11}^{12} \frac{12!}{(12-x)!\,x!}(0.9)^x(1-0.9)^{12-x} = 0.658$$

and $C = 1 - \Pr \geq X\,(3) = 1 - 0.658 = 0.34.2$

In this case there is a 34.2 % confidence level that the breaker has a 90 % probability of withstanding the lighting impulse.

7.2.3.4 IEC Bias test

In recognition of the fact that a lighting stroke may reach the circuit breaker at any time and of the effects produced by the impulse wave upon the power frequency wave, IEC requires that for all circuit breakers rated 300 kV and above an impulse bias test be performed. This tests is made by applying a power frequency test voltage to ground which when measured in relation to the peak of the impulse wave is no less than 0.572 times the rated voltage of the circuit breaker. No equivalent requirement has been established by ANSI.

7.2.3.5 Switching Impulse Withstand

This requirement is applicable to circuit breakers rated, by ANSI at 362 kV, or above and by IEC at 300 kV and above. The reason why these requirements are specified only at these values is because, at the lower voltage ratings, the peak value of the specified power frequency withstand voltage exceeds the 3.0 per unit voltage surge which is the value that has been selected as the maximum uncontrolled switching surge that may be encountered on a system.

At the voltage levels, where the switching impulse voltage is specified, there are two levels listed and with the exception of the across the terminals rating for a 362 kV circuit breaker, all the other values are lower than the 3.0 p.u. value mentioned earlier. The lower voltages have been justified by arguing that, when the circuit breaker is closed, the breaker is inherently protected by the source side arresters which are always used. However, with the breaker in the open position, unless there are line side arresters, the breaker would be unprotected and the specified levels would be inadequate, but nevertheless, it is also recognized that all circuit breakers at these voltage levels use some form of surge control. This practice is reflected in the factors, that are listed on ANSI C37.06, for circuit breakers specifically designed to control line closing switching surge maximum voltages.

7.2.4 Rated Transient Recovery Voltage

Recalling what was said earlier, in chapter 3, the transient recovery voltages that can be encountered in a system can be rather complex and difficult to calculate without the aid of a digital computer. Furthermore, we learned that the TRV is strongly dependent upon the type of fault being interrupted, the configuration of the system and the characteristics of its components, i.e. transformers, reactors, capacitors, cables, etc.

We had also made a distinction for terminal faults, short line faults and for the initial TRV condition. The standards recognize all of these situations and consequently have specific requirements for each one of these conditions. As stated earlier, for standardization and testing purposes, the prospective, or inherent TRV could be simplified so it could be described in terms of simpler waveform envelopes. It is important to emphasize that what follows always refers to the inherent TRV, that means the recovery voltage produced by the system alone discounting any modifications or any other distortions that may be produced by the interaction of the circuit breaker.

7.2.4.1 Terminal faults

Based on the above, ANSI had adopted two basic waveforms to simulate the most likely envelopes of the TRV for a terminal fault condition under what can be considered as generic conditions.

For breakers rated 72.5 kV and below the waveform is mathematically defined as:

$$E_{TRV} = 1 - \text{cosine } \omega t$$

where:

$$\omega = \frac{\pi}{T_2} \quad \text{and}$$

T_2 = Specified rated time to voltage peak

For breakers rated 121 kV and above the wave form is approximately defined as the envelope of the combined Exponential-Cosine functions, and the exponential portion [6] is given by the following equation

$$e = E_1 \left[1 - \varepsilon^{-\alpha t} \left(\cosh \beta t + \frac{\alpha}{\beta} \sinh \beta t \right) \right]$$

where:

$$\alpha = \frac{1}{2ZC}$$

$$\beta = \sqrt{\alpha^2 - \frac{1}{LC}}$$

$$Z = \frac{R \times 10^6}{\sqrt{2} \, \omega I} \quad \text{Ohms}$$

$$\omega = \frac{\pi}{T_2}$$

$$C = \frac{T_1 \times 10^{-6}}{Z} \quad \text{Farads}$$

$$L = \frac{\text{Rated Max. Voltage}}{\sqrt{2} \, \omega I}$$

and where E_1, E_2, R, I, T_1, and T_2 are the values given in ANSI C37.06.

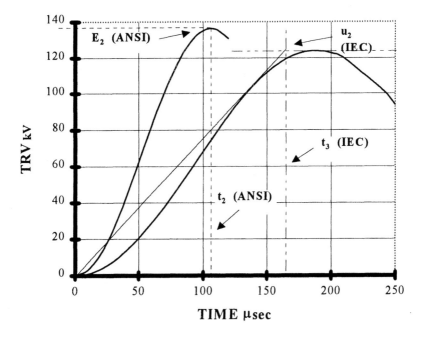

Figure 7.4 Comparison of ANSI (1– cos) and IEC Two Parameter TRV for 72.5 kV Rated Circuit Breaker.

IEC also defines two TRV envelopes, one that is applicable to circuit breakers rated 72.5 kV and below and which is defined by what is known as the Two Parameter Method, and the second which is applicable to circuit breakers rated above 72.5 kV and which is known as the Four Parameter Method. In figures 7.4 and 7.5 the TRV envelopes as defined by ANSI and IEC are compared, and the curves shown correspond to a 72.5 kV and a 145 kV rated circuit breakers respectively.

The method employed by IEC assumes that the TRV wave may be defined by means of an envelope which is made of three line segments, however when the TRV approaches the 1-Cosine, or the damped oscillation shape the envelope resolves itself into two segments.

The following procedure for drawing the aforementioned segments as described in [7] can be used.: The first line segment is drawn from the origin and in a position that is tangential to the TRV without crossing this curve at any point. The second segment is a horizontal line which is tangent the highest point on the TRV wave. The third segment is a line that joins the previous two

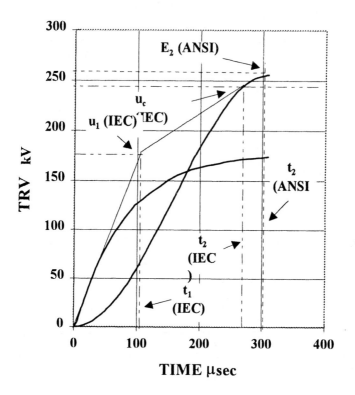

Figure 7.5 TRV Comparison of ANSI (Exp.-Cos) and IEC Four parameter envelopes for a 145 kV Circuit breaker.

segments, and which is drawn tangent to the TRV wave without crossing this wave at any point. However, when the point of contact of the first line and the highest peak are comparatively close to each other, the third line segment is omitted and the two parameter representation is obtained.

When all three segments are drawn the four parameters defined are:

u_1 = Intersection point of first and third line segments corresponds to the first reference voltage in kilovolts.

t_1 = time to reach u_1 in microseconds.

u_c = Intersection point of second and third line segments corresponds to the second reference voltage (TRV peak) in kilovolts.

t_2 = time to reach u_c in microseconds.

The two parameter line is defined by:

u_c = corresponds to reference voltage (TRV peak) in kilovolts.

t_3 = time to reach u_c in microseconds.

Not only the adopted waveforms are different but, there are a number of additional differences between the transient recovery voltages that are specified by ANSI and IEC; for example, the TRV characteristics described by IEC are independent of the interrupting current ratings of the breaker, while the ANSI characteristics vary in relation to the interrupting rating. ANSI specifies different rates of change, (dv/dt), for the TRV. These rates are also dependent on the interrupting current ratings but, IEC establishes only a constant equivalent rate that is applicable to all ratings. Finally, ANSI assumes that all systems will be ungrounded, which is a conservative approach because it gives a 1.5 factor that is used as the multiplier for the first phase to clear the fault. IEC, in the other hand, makes a distinction between grounded and ungrounded applications and accounts for these differences by the usage of the multiplier 1.3, or 1.5 respectively for the first phase to clear. However, even in those cases where the same 1.5 multiplier is used by both standards, the corresponding ANSI values are higher because, based on the recommendations made by the AEIC study committee [8] the amplitude factors used. are equal to 1.54 for circuit breakers rated 72.5 kV and below and 1.43 for circuit breakers rated above 72.5 kV. In contrast, the amplitude factor used by IEC is 1.4 for all voltage ratings.

There is a relatively significant time differences observed between the ANSI and the IEC time to crest requirements for circuit breakers rated below 72.5 kV. This difference can be explained by considering the European practice where most of the circuits at these voltages are fed by cables, and consequently, due to the inherent cable's capacitance, the natural frequency of the response is lowered.

The influence of capacitance, on the source side of the breaker, produces a slower rate of rise of the TRV and therefore, what amounts to a delay time before the rise in the recovery voltage is observed. ANSI does not specify a delay for the conditions of a 1- cosine envelope because in this case the initial rate is equal to zero, IEC however, due to the use of the two parameter envelope, does specify a time delay which ranges; from a maximum of 16 microseconds, down to 8 microseconds. For circuit breakers rated above 121 kV the delay time specified by IEC is a constant with a value of 2 microseconds, ANSI specifies different delay times for different voltage ratings and these times range from a low of 2.9 microseconds at 121 kV to a maximum of 7.9 microseconds for 800 kV circuit breakers.

In both standards, consideration is given to the condition where currents significantly lower than the rated fault currents, and usually in the approximate range of 10 to 60 %. are interrupted. This condition is assumed to occur when the fault occurs in the secondary side of the transformer, in such cases the

Table 7.3
Breakers Rated 72.5 kV and below
Multiplying Factors for Reduced Fault Currents

ANSI			IEC		
Fault Current %	E_2	t_2	Fault Current %	u_c	t_3
60	1.07	0.67	60	1.07	0.43
30	1.13	0.40	30	1.07	0.21
10	1.17	0.40	10	1.07	0.21

Table 7.4
Breakers Rated 121 kV and above
Multiplying Factors for Reduced Fault Currents

ANSI			IEC			
Fault Current %	E_2	t_2	Fault Current %	u_c	t_3	
60	1.07	0.50	60	1.07	1.00	
30	1.13	0.20	30	1.07	1.00	
10	1.17	0.20	10	1.09	121 kV	0.20
					145 kV to 245 kV	0.17
					362 kV	0.14
					550 kV	0.12
					800 kV	0.09

currents are reduced due to the higher reactance of the transformer. These higher reactance in turn change the natural frequencies of the inherent TRV. To account for these changes ANSI and IEC have selected three levels of reduced currents where the specified peak voltage is higher and the time to reach the peak is reduced. The factors by which, the voltages are increased and the times are reduced, are tabulated in Tables 7.3 and 7.4. The net result is that the peak TRV values given by ANSI are higher than the ones required by IEC but, the time required to reach the peak is always higher according to IEC.

Table 7.5
Comparison of SLF Parameters

	ANSI	IEC
Surge Impedance Z Ohms	All = 450	All = 450
Amplitude Factor d	All = 1.6	All = 1.6
Time delay (seconds	0	(170 kV = 0.2 (245 = 0.5

7.2.4.2 Short Line Faults

An important difference between ANSI and IEC for short line faults is that IEC requires this capability only on circuit breakers rated 52 kV and above, and which are designed for direct connection to overhead lines. ANSI requires the same capability for all outdoor circuit breakers. Other differences found between the two standards are compared in Table 7.5.

7.3 Current Related Ratings

7.3.1 Rated Continuous Current

The continuous current rating is that which set the limits for the circuit breaker temperature rise. These limits are chosen so that a temperature run away condition, as it was described in the previous chapter, is avoided when is taken into consideration the type of material used in the contacts, or conducting joints, and secondly, so that the temperature of the conducting parts, which are in contact with insulating materials, do not exceed the softening temperature of such material. The temperature limits are given in terms of both, the total temperature and the temperature rise over the maximum allowable rated ambient operating temperature. The temperature rise value is given to simplify the testing of the circuit breaker because as long as the ambient temperature is between the range of 10 to 40 °C no correction factors need to be applied.

The preferred continuous ratings specified by ANSI are, 600, 1200, 1600, 2000, or 2000 Amperes. The corresponding IEC ratings are based on the R10 series of preferred numbers and they are; 630, 800, 1250, 1600, 2000, 3150, or 4000 Amperes.

The choices of the numerical values for the continuous currents, made by each of the two standards, are not that different, but the real significance of the ratings is the associated maximum allowable temperature limits that have been established. These temperature limits, as they are currently specified, by each of the standards, are shown in Tables 7.6 and 7.7.

The maximum temperature limits for contacts and for conducting joints are chosen based on the knowledge of the relationships that exist for a given material between the change in resistance due to the formation of oxide films, and the prevailing temperature at the point of contact. For insulating materials the temperature limits are well known and are directly related to the mechanical characteristics of the material. These limits follow the guidelines that are given by other standards, such as ASTM, that have classified the materials in readily identifiable groups.

An additional requirement given is the maximum allowable temperature of circuit breaker parts that may be handled by an operator, these parts are limited to a maximum total temperature of 50 °C and for those points that can be accessible to personnel the limit is 70 °C. But, in any case external surfaces are limited to 100 °C.

7.3.1.1 Preferred Number Series

Preferred numbers are series of numbers that are selected for standardization purposes in preference of any other numbers. Their use leads to simplified practices and lead to reduced number of variations.

The preferred numbers are independent of any measurement system and therefore they are dimensionless. The numbers are rounded values of the following five geometric series of numbers: $10^{N/5}$, $10^{N/10}$, $10^{N/20}$, $10^{N/40}$ and $10^{N/80}$, where N is an integer in the series 0, 1, 2, 3, etc. The designations used for the five series are R5, R10, R20, R40 and R80 respectably, where R stands for Renard of Charles Renard, the originator of this series, and the number indicates the root of ten on which the series is based.

The R10 series is frequently used to establish current ratings. This particular series gives 10 numbers that are approximately 25 % apart. These numbers are: 1.0, 1.25, 1.60, 2.0, 3.15, 4.00, 5.00, 6.30, and 8.00, this series can be expanded by using multiples of 10.

7.3.2 Rated Short Circuit Current

The short circuit current rating as specified by both standards corresponds to the maximum value of the symmetrical current that can be interrupted by a

Table 7.6
IEC 294 Temperature Limits

COMPONENT		MAXIMUM TEMPERATURES	
CONTACTS		Total Temp°C	Temp. Rise°C
Bare Copper	In Air	75	35
	In SF$_6$	90	50
	In Oil	80	40
Silver or Nickel Plated	In Air	105	65
	In SF$_6$	105	65
	In Oil	90	50
Tin Plated	In Air	90	50
	In SF$_6$	90	50
	In Oil	90	50
CONNECTIONS			
Bare Copper	In Air	90	50
	In SF$_6$	105	65
	In Oil	100	60
Silver or Nickel Plated	In Air	115	75
	In SF$_6$	115	75
	In Oil	100	60
Tin Plated	In Air	105	65
	In SF$_6$	105	65
	In Oil	100	60
EXTERNAL TERMINALS to CONDUCTORS			
Bare		90	50
Silver, Nickel, or Tin plated		105	65
INSULATING MATERIALS			
Class Y		90	50
Class A		100	60
Class E		120	80
Class B		130	90
Class F		155	115
Class H		180	140

Table 7.7

ANSI C37.04 Temperature Limits

COMPONENT		MAXIMUM TEMPERATURES	
CONTACTS		Total Temp°C	Temp. Rise °C
Copper		70	30
Silver Alloy or Equivalent	In Air In Oil	105 90	65 50
CONNECTIONS			
Copper		70	30
Silver Plated or Equivalent	In Air In Oil	105 90	65 50
EXTERNAL TERMINALS to CONDUCTORS			
Bare		85	45
INSULATING MATERIALS			
Class O		90	50
Class A		100	60
Class B		130	90
Class F		155	115
Class H		180	140
Class C		220	180
Oil (top oil, upper layer)		90	50

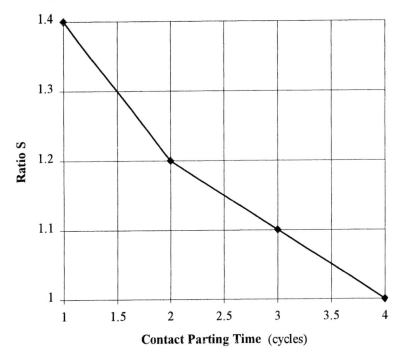

Figure 7.6 Factor S, Ratio of Symmetric to Asymmetric Interrupting Capabilities.

circuit breaker. The values of these currents for outdoor breakers are based on the R10 series.

Associated with the symmetrical current value, which is the basis of the rating, there are a number of related capabilities, as they are referred by ANSI, or as definite ratings as specified by IEC. The terminology used for these capabilities may differ, but the significance of the parameters is the same in both standards. What it is important is to realize that both standards use the same assumed X/R value of 17 as the time constant which defines most of the of the related requirements for all transient current conditions.

7.3.3 Asymmetrical Currents

In most cases, as described in chapter 2, the ac short circuit current has an additional dc component, which is generally referred as the "per cent dc component of the short circuit current." The magnitude of this component is a function of the time constant of the circuit, (X/R = 17), and of the elapsed time between the initiation of the fault and the separation of the circuit breaker contacts.

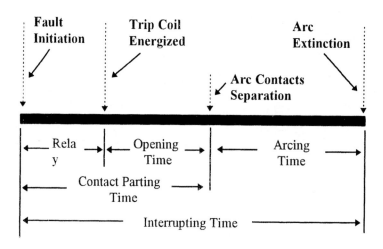

Figure 7.7 Interrupting Time Relationships

ANSI establishes an asymmetry factor, S which, when multiplied by the symmetrical current, defines the asymmetrical value of the current. The actual value of the factor S can be calculated using the following relationship

$$S = \sqrt{1 + 2\left(\frac{\%dc}{100}\right)^2}$$

where the % *dc* equals the dc component of the short circuit current at the time of contact separation

ANSI however, to simplify the process, and as illustrated in figure 7.6, has specified definite values for S, and where they are referred to the rated interrupting time of the circuit breaker. The specified factors are: 1.4, 1.3, 1.2, 1.1 and 1.0 for rated interrupting times of 1, 2, 3, 5, and 8 cycles respectively. IEC establishes the magnitude of the dc component based on a time interval consisting of the sum of the actual contact opening time plus one half cycle of rated frequency. In essence, there is no difference between the requirements of the two standards, except that in the ANSI way the factors are not exact numbers but approximations for certain ranges of asymmetry.

Interrupting Time. The interrupting time of the circuit breaker consists of the summation of one-half cycle of relay time, plus the contact opening time, which is the time that it takes the breaker, from the instant the trip coil is energized until the time where the contacts separate, plus the maximum arcing time of the circuit breaker.

Note that the contact parting time is the summation of the relay time plus the opening time. These relationships are illustrated in figure 7.7.

7.3.4 Close and Latch, or Peak Closing Current

The peak asymmetrical closing current is also referred as the close and latch current or the peak making current. This current rating is established for the purpose of defining the mechanical capability of the circuit breaker, its contacts and its mechanism to withstand the maximum electromagnetic forces generated by this current. The magnitude of this current is expressed in terms of multiples of the rated symmetrical short circuit current. ANSI used to specify a 2.7 factor, but recently it was reduced to 2.6 for the sake of mathematical accuracy. IEC specifies a 2.5 multiplier. The difference between the two values is due to the difference in the rated power frequencies.

The equation that is used to calculate the peak current is:

$$I_C = \sqrt{2}\, I \left[(1 - \cos\omega t) + \varepsilon^{-\frac{t}{\tau}} \right]$$

where:

I_C = Peak of making current

I = Symmetrical short circuit current

$t = 0.4194$ cycles = elapsed time to current peak

$\tau = \dfrac{X}{R\omega} = \dfrac{17}{377}$ (for 60 Hz) or $\dfrac{17}{314}$ (for 50 Hz)

7.3.5 Short Time Current

The purpose of this requirement is to assure that the short time heating capability of the conducting parts of the circuit breaker are not exceeded. By definition the short time current rating, or related capability, is the rms. value of the current that the circuit breaker must carry, in the closed position, for a prescribed length of time.

The magnitude of the current is equal to the rated symmetrical short circuit current that is assigned to that particular circuit breaker and the required length of time is specified as 3 seconds by ANSI and as 1 second by IEC. Nevertheless, IEC recommends a value of 3 seconds if longer than 1 second periods are required. The 1 second specification of IEC corresponds to their allowable tripping delay which in the IEC document is referred to as the rated duration of short circuit.

Even though ANSI requires a three (3) seconds withstand, the maximum allowable tripping delay is specified as two (2) seconds for indoor circuit breakers and for outdoor circuit breakers that have a rating of 72.5 kV or less, and for circuit breakers with voltage ratings at, or above 121 kV, the time requirement is one (1) second, which also happens to be the time specified by IEC.

Recognizing, that the duration of the short time current does not have to be any greater than the maximum delay time that is permitted on a system, ANSI is in the process of adopting the shorter time requirements.

7.3.6 Rated Operating Duty Cycle

The rated operating duty cycle as is referred in ANSI, is known as the rated operating sequence by IEC.

ANSI specifies the standard operating duty as a sequence consisting of the following operations; CO—15 s—CO, that is, a close operation followed by an immediate opening and then, after a 15 second delay another close-open operation.

IEC offers two alternatives, one is O -3 min.—CO-3 min.—CO and the second alternative is the same duty cycle prescribed by ANSI. However, for circuit breakers that are rated for rapid reclosing duties the time between the opening and close operation is reduced to 0.3 seconds.

ANSI normally refers to the reclosing duty as being an open—0 sec.—close-open cycle which implies that there is no time delay between the opening and the closing operation. However in C37.06 the rated reclosing time, which corresponds to the mechanical resetting time of the mechanism, is given as 20 or 30 cycles which corresponds to 0.33 or 0.5 seconds respectively depending upon the rating of the circuit breaker, the reference to 0 seconds is supposed to mean that no intentional external time delays can be included.

When referring to reclosing duties ANSI allows a current derating factor R to be applied, but IEC does not make any derating allowance. In reality the derating factor was only applicable to older interrupting technologies and with modern circuit breakers there is no need for any derating.

Have you ever noticed the lights to go out, come back instantly, go out again and then after a short period of time, (15 seconds), come on again and if they go out again there is a long period before power is restored.? Well, what you have witnessed is a circuit breaker duty cycle including the fast reclosing option.

7.3.7 Service Capability

The service capability is a requirement that is only found in ANSI, what it defines is a minimum acceptable number of times that a circuit breaker must interrupt its rated short circuit current without having to replace its contacts. This capability is expressed in terms of the accumulated interrupted current and for older technologies (oil and air magnetic circuit breakers) a value of 400 % is specified, for modern SF_6 and vacuum circuit breakers the required accumulated value is 800 % of the maximum rated short circuit current.

7.4 Additional Switching Duties

Aside from switching short circuit currents we know that circuit breakers must also execute other types of switching operations. The requirements for these operations are defined in the corresponding standards but, as is the case with most of all other requirements, there are some noteworthy differences which are discussed in the sections that follow.

7.4.1 Capacitance Switching

All of the conditions that are considered to be related to capacitance switching duties are listed separately below.

7.4.1.1 Single Bank

ANSI mandates that all circuit breakers must be designed to meet the requirements for the "General-Purpose Circuit Breaker" classification as is listed in ANSI C37.06; meanwhile IEC lists this duty as an optional rating, it does not assign any interrupting current values and only makes the recommendation that these values be selected using the R 10 series.

The values of currents for switching single capacitor banks that are applicable to general purpose breakers are pre-assigned by ANSI and they have been selected using the R10 series

7.4.1.2 Back to Back Capacitor Banks

Again this is an optional rating according to the IEC standard and in line with their practice, regarding single capacitor banks, the interrupting current values are not specified. ANSI handles the back-to-back capacitor bank rating by defining a group of "Definite-Purpose Circuit Breakers". This is an optional rating; and it is not expected that every circuit breaker would meet this requirement, unless it is specifically designed for this purpose. In reality, however, most circuit breakers involving modern technologies are normally designed with this requirement in mind, and most manufacturers produce only one version rather than having two different designs.

7.4.2 Line Charging

Line charging requirements are included by ANSI as part of their capacitance switching specifications and therefore all circuit breakers indoor and outdoor have certain specific assigned requirements. In IEC standards the rating for line charging is only applicable to circuit breakers which are intended for switching overhead lines and rated at 72.5 kV or above. There is also differences on the magnitude of the specified currents. At 72.5 kV and up to 170 kV IEC values are approximately 30 to 50 % lower than the ANSI values for general purpose circuit breakers, and about 90 % lower than those for definite purpose breakers. At the 245 and 362 kV circuit breakers levels the IEC specified line charging current ratings are approximately 25 % higher than the ANSI currents for general purpose breakers but are about 30 % and 60 % lower than ANSI values for the respective breakers. At 550 and 800 kV the required line charging currents are the same in both standards.

7.4.3 Cable Charging

Cable charging is considered to be only a special case of capacitance switching, and therefore is included, by ANSI with all of the other requirements for capacitance switching. Since IEC does not make capacitance switching a mandatory requirement, then neither is cable switching in their standard

7.4.4 Reignitions, Restrikes, and Overvoltages

These requirements serve to define what is considered to be an acceptable performance of the circuit breaker during capacitive current switching. The limitations that have been specified in the standards are aimed to assure that the effects of restrikes and the potential for voltage escalation, which have been described earlier, are maintained within safe limits.

The maximum overvoltage factor is specified because it is recognized that certain types of circuit breakers, most notably oil types, are prone to have restrikes. What this rating does is to allow restrikes, only when appropriate means have been implemented within the circuit breaker to limit the overvoltages to the maximum value given in the respective standard. The most common method used for voltage control is the inclusion which can be the use of shunt resistors. to control the magnitude of the overvoltage.

REFERENCES

1. ANSI C84.1-1982, Voltage Ratings for Electric Power Systems and Equipment (60 Hz).
2. ANSI C92.2-1974 Preferred Voltage Ratings for Alternating-Current Electrical Systems and Equipment Operating at Voltage above 230 kilovolts Nominal.
3. L. Wagner, J. M. Clayton, C. L. Rudasill, and F. S. Young, Insulation Levels for VEPCO 500-kV Substation Equipment, IEEE Transactions Power Apparatus and Systems, Vol. PAS. 83: 236-241, March, 1964.
4. L. Wagner, A. R. Hileman, Lightning Surge Voltages in Substations Caused by Line Flashovers, AIEE Conference Paper CP 61-452, Presented at the Winter General Meeting, New York Jan./Feb. 1961.

5. ANSI/IEEE C37.04-1979, Rating Structure for AC High-Voltage Circuit Breakers Rated on a Symmetrical Current Basis.
6. IEC-56 1987, High Voltage Alternating Current Circuit Breakers.
7. Transient Recovery Voltages on Power Systems (as Related to Circuit Breaker Performance), Association of Edison Illuminating Companies, New York, 1963.
8. NEMA SG-4 -1995 Alternating Current High Voltage Circuit Breakers.
9. IEEE 1-1969, General Principles for Temperature Limits in the Rating of Electric Equipment.
10. IEEE 4-1978 Standard Techniques for High Voltage Testing.

8

SHORT CIRCUIT TESTING

8.0 Introduction

Since a circuit breaker represents the last line of defense for the whole electric system it is imperative to have a high degree of confidence in its performance, and this confidence level can only be attained by years of operating experience, or by extensive testing under conditions that simulate those that are encountered in the field applications. Short circuit testing, whether it involves an individual interrupter, or a complete circuit breaker, is one of the most essential and complex tasks that is performed during the development process.

Short circuit testing is also important from the technical engineering design point of view because, in spite of all the knowledge gained about circuit interruption and today's ability to model the process, the modeling itself is based, to a great extent, on the experimental findings and consequently, testing becomes the fundamental tool that is used for the development of circuit breakers.

Short circuit testing has always presented a challenge, because what must be replicated is the interaction of a mechanical device, the circuit breaker, and the electric system, where as we already know, the switching conditions can vary quite widely depending upon the system configuration and therefore so can the type of conditions which must be demonstrated by tests.

Another challenge has always been the development of appropriate test methods that can overcome the potential lack of sufficient available power at the test facility. One has to consider that the test laboratory should basically be able to supply the same short circuit capacity as that of the system for which the circuit breaker that is being tested is designed, however, this is not always possible, specially at the upper end of the power ratings.

Presently it is possible to test a three phase circuit breaker up to 145 kV and a maximum symmetrical current of 31.5 kA. Anything above these levels must be tested on a single phase basis, unless it is tested directly out of an electrical network, a condition which is highly unlikely.

In the early days of the industry, however, practically all testing was done in the field using the actual networks to supply the required power. Even today, in some cases, direct field tests are still being performed. However, in the majority of the cases the testing is done at any of a number of dedicated testing stations available in many countries around the world. The majority of these test stations use their own power generators which have been specially designed for short circuit testing.

With a three phase capacity of 8,400 MVA, N. V. KEMA, Arnhem, The Netherlands, is the largest test station in the world. An aerial view of this test laboratory is shown in figure 8.1. Its subsidiary, KEMA-Powertest located in Chalfont, Pennsylvania has the largest capability (3,250 MVA) in the US.

In addition to the above named laboratories there are a number of other important testing facilities around the world; in North America the newest laboratory with a three phase capacity of 2,200 MVA is LAPEM which is located in Irapuato, Mexico, a view of this modern and well equipped facility is shown in figure 8.2.

Additionally, in the US there is PSM in Pennsylvania, and in Canada there is IREQ in Quebec, and Power-Tech in Vancouver. In Europe there are Siemens and AEG in Germany, CESI in Italy, Electricity de France in Fonteney, and ABB in both Sweden and Switzerland, MOSKVA in Russia, VUSE in Czechoslovakia, Warszava in Poland, and in Japan Mitsubishi and Toshiba laboratories.

Figure 8.1 Aerial view of the N. V. KEMA laboratory located in Arnhem, The Netherlands (Courtesy of B.V. KEMA, Arnhem, The Netherlands).

Figure 8.2 Aerial view of the LAPEM laboratory located in Irapuato, Gto., Mexico (Courtesy of LAPEM, Irapuato, Gto. Mexico).

8.1 Test Methodology

A great deal of latitude is given as far as short circuit current design testing of a circuit breaker is concerned. Because of the levels of power needed, the complexity of the tests and the very high costs involved with these test, this latitude is not only a convenience but a necessity.

To reduce the otherwise required amount of testing, it is permissible to analyze results from similar design tests and to use engineering judgment to evaluate these results. However, this judgment must be technically sound, supported by good data and backed-up by a strong knowledge about the characteristics of the circuit breaker in question.

As long as sufficient evidence is gathered, and as long as it is satisfactorily demonstrated that the most severe testing conditions are met, one can certify the interrupter's performance by combining, in any order, the listed operating test duties. The tests do not necessarily have to be performed in any particular sequence, and that they do not even have to be done using the same interrupter, as long as the total accumulated current duty is reached with one individual unit.

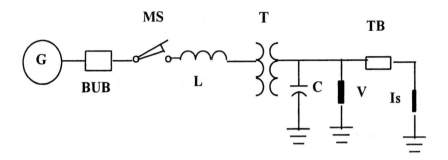

Figure 8.3 Elementary schematic diagram of the basic circuit used for short circuit current tests.

Since, even at the largest of the test stations, it is not always possible to supply the full amount of power needed, alternate, appropriate methods that require less power but, that yield equivalent results had to be developed. Other challenges that had to be overcome had to do principally with signal isolation, surge protection, reliable control, and high speed instrumentation.

The difficulties, that can be encountered during high current tests, can only be appreciated when one considers the need for an absolutely proper sequential coordination of a number of events, and where a failure within the sequence may result in aborting the test or what is worse in a disastrous failure, and where, during an individual test, there is only a single chance to capture the test record.

In recent years, with the advent of high speed, high resolution, digital instrumentation and data acquisition systems the quality and reliability of tests, test records, and in general of the total accompanying documentation has been greatly enhanced. Today it is possible to measure, store and later replay, at high resolution levels the complete test sequence, this provides a very powerful tool for the analysis of the events taking place during current interruption.

A typical short circuit current test set-up, is illustrated by the schematic diagram of figure 8.3. The test circuit consists of a power source (G) which can be either, a specially designed short circuit generator, such as the ones that are illustrated in figures 8.4 and 8.5, or the system's electric network itself.

For the protection of the generator, or power source, a high capacity back-up circuit breaker (BUB) is used for interrupting the test current, in the event that the circuit breaker being tested (TB) would fail to interrupt the current. Back-up circuit breakers, in the majority of cases, are of the air blast type. and a single pole of the air blast back-up circuit breakers manufactured by AEG and which are used at LAPEM is shown in figure 8.6.

Figure 8.4 Generator hall with 2,250 MVA and 1,000 MVA short-circuit test generators. (Courtesy KEMA-Powertest, Chalfont, PA, USA.)

Figure 8.5 View of LAPEM's Test Generator (Courtesy of LAPEM, Irapuato, Gto. Mexico)

Figure 8.6 Single pole of an air-blast high current interrupting capacity circuit breaker used for back-up protection of test generator at LAPEM. (Courtesy of LA-PEM, Irapuato, Gto. Mexico.)

In series with the back-up circuit breaker there is a high speed making switch (MS) which is normally a synchronized switch capable of independent pole operation and of precise control for closing the contacts at an specific point on the current wave. This type of operation allows precise control for the initiation of the test current and consequently gives the desired asymmetry needed to meet the specific conditions of the test.

Current limiting reactors (L) are connected in series with the making switch, their mission is to limit the magnitude of the test current to its required value. These reactors can be combined in a number of different connection schemes to provide with a wide range of impedance values.

Specially designed test transformers (T), such as those shown in figure 8.7, that have a wide range of variable ratios are connected between the test circuit breaker and the power source. These transformers are used primarily to allow for flexibility for testing at different voltage levels, and in addition to provide isolation between the test generator and the test piece.

Figure 8.7 Test transformers installed at the LAPEM laboratory. Two transformers per phase are used in this installation. (Courtesy of LAPEM, Irapuato, Gto., Mexico.)

Across the test breaker terminals a bank of TRV shaping capacitors (C) is connected and to measure these voltages at least one set of voltage dividers (V) which are usually of a capacitive type are used. In most cases the short circuit current flowing through the test device is measured by means of a shunt (Is) however, measurements using current transformers are also often made, specially for the current at the upstream side of the test piece.

When one is interested in investigating the interrupting phenomena at the precise instant of a current zero, the use of coaxial shunts is highly recommended for distortion free measurements; however, caution should be taken to limit the current flowing through the shunt since these type of shunts normally have only limited current carrying capabilities.

The typical test set-up that has just been described is used for practically all direct tests as the primary current source, and even for those situations where the available power is insufficient and where alternate test methods have been developed, the primary source of current still is similar to the one that has been just described.

8.1.1 Direct Tests

A direct test method is one where a three phase circuit breaker is tested, on a three phase system, and at a short circuit MVA level equal to its full rating. In other words this is a test where a three phase circuit breaker is tested on a three phase circuit at full current and at full voltage. It should be obvious that testing a circuit breaker under the same conditions at which it is going to be applied is the ultimate demonstration for its capability and naturally, whenever possible, this should be the preferred method of test.

8.1.2 Indirect Tests

Indirect tests are those which permit the use of alternate test methods to demonstrate the capabilities of a circuit breaker for applications in three phase grounded or ungrounded systems.

The methods most commonly employed are:

1. Single phase tests
2. Two part tests
3. Unit tests
4. Synthetic tests

8.1.3 Single Phase Tests

When one thinks only in terms of an individual interrupter it is realized that, as far as the interrupter is concerned, it makes no difference whether a three phase or a single phase power source is used, as long as the current and the recovery voltage requirements for the test are fulfilled, therefore, the use of a single phase test procedure is totally acceptable; however, when the final application of the interrupter is considered, and since in most cases it turns up to be in a three phase circuit breaker, then the neutral shift of the source voltage and some of the potential mechanical interactions that may occur between the poles have to be taken into consideration.

In the first place, in a three phase application at the instant of current zero, and in the phase where the current interruption is about to take place, the interrupter itself does not know that there are another two phases lagging slightly behind on time. If the first phase which sees a current zero fails to interrupt, then the next sequential phase will attempt to clear the circuit. This, basically, gives the circuit breaker an additional two chances to interrupt the current and therefore, as it has been shown by W. Wilson [1], there is a higher probability of successfully interrupting a three phase current than of interrupting a single phase fault.

The oscillogram which is shown in figure 8.8 depicts the condition where interruption is attempted sequentially at each current zero. As it can be seen in the figure, the first attempt is made on phase B (shown as B_1), since no interruption is accomplished the second attempt is made at the next current zero which is in phase A (shown as A_1), again the interrupter is not successful on this try and finally on the third try, the current is interrupted in phase C. The other two phases are seen to interrupt the current simultaneously at A-B.

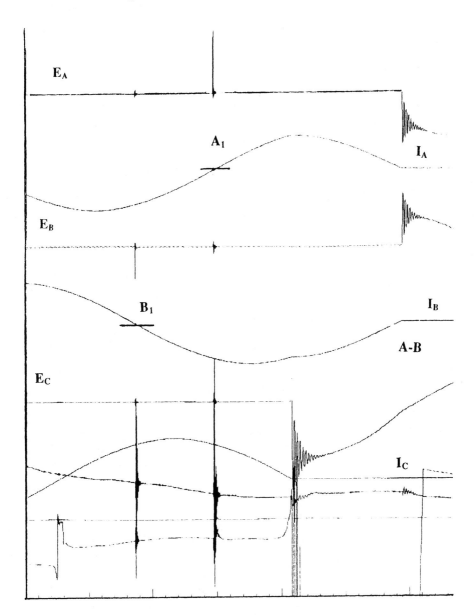

Figure 8.8 Oscillogram of a three phase asymmetrical current interruption test, It illustrates the sequential attempts that are made by the interrupter to clear the current at each successive current zero of each of the three phases.

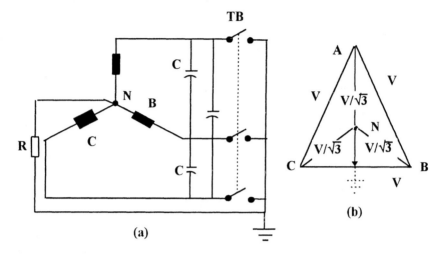

Figure 8.9 Source voltage shift following the interruption of current by one phase (first phase to clear).

During the first few microseconds following the interruption of the current it only matters that the proper transient recovery voltage be applied across the interrupter, and since in a three phase circuit, as the high frequency oscillations of the load side TRV die down, and before the other two phases interrupt the current, the source side power frequency recovery voltage is reduced to 87 % of the line to line voltage, due to the neutral shift, as seen in the vector diagram for the power source voltage which is shown in figure 8.9.

After the currents in all the phases are interrupted, the voltage in each phase becomes equal to the line to neutral voltage, which corresponds to 58 % of the line to line voltage. However, this reduction to 58% of the voltage occurs approximately four milliseconds after the current is interrupted by the first phase, this is a relatively long time, especially with today's interrupters, and when taken into consideration that it is long enough, after the interrupter has withstood the maximum peak of the TRV then it is justifiable to expect that the interrupter has regained its full dielectric capability and thus, in most cases it becomes only academic the fact that the voltage reduction takes place.

Aside from the purely electrical considerations that have been given above, the possible influence of the electromechanical forces produced by the currents and of the gas exhaust from adjacent poles, should be carefully evaluated. It is also important to carefully balance the energy output of the operating mechanism, to compensate for the reduced operating force needed to operate a single pole, so that the proper contact speeds are attained. This last recommendation is specially important when testing puffer type circuit breakers.

Naturally, these concerns about the possible pole interaction do not apply to those circuit breakers which have independently operated poles.

8.1.4 Unit Tests

Unit tests can be considered to be simply a variation of the single phase test method which has been used almost exclusively for the extra high voltage class of circuit breakers where, as it should be recalled, it is common practice to install several identical interrupters in series on each pole mainly for the purpose of increasing the overall voltage capability of the circuit breaker.

This test method demonstrates the interrupting capability of a single interrupter from a multiple interrupter pole, provided that they are identical interrupters. The test is performed at full rated short circuit current and at a voltage level that is equivalent to the ratio of the number of interrupters, used in the pole assembly, to the full rated voltage of the complete pole and where the distributed voltage is properly adjusted to compensate for the uneven voltage distribution that normally exists across each series interrupter unit and which is due to the influence of stray capacitance, adjacent poles, and the proximity and location of the ground planes. However, in any case the test voltage must be at least equal to the highest stressed unit in the complete breaker.

When this test method is used the frequency of the TRV does not change but, due to the lower voltage peak of the individual interrupter, the rate of rise of the recovery voltage is proportionally lower. This characteristic response holds not only for terminal fault tests but also for short line fault tests.

Whenever the unit test method is utilized and, as it was the case with the previous test method, care must be taken to properly scale the mechanical operating parameters to ensure the validity of the tests.

8.1.5 Two Part Tests

A two part test consists of two essentially independent tests. The first test, is one where the interrupter is tested at full rated voltage and at a reduced current. In the second test the maximum current is applied at a reduced voltage.

The idea behind this method is to test for the dielectric recovery region with the first portion of the test and then to complement the results by exploring the thermal recovery region by means of the second test. Application of this test method has always been limited to the extra high voltage interrupters, where, as it should be recalled, the TRV is represented by a waveform that is composed of an exponential and a (1-cos) function. When the two part tests are performed, the first portion of the test is made at full rated current and with a TRV that is equal to the exponential portion of the waveform. The second part of the test is made at a reduced current but at full voltage and with a TRV equal to the (1-cosine) component and where the requirements for the voltage peak and for the time to reach this peak are verified.

This test method is often difficult to correlate with actual operating conditions and therefore it is somewhat difficult to justify. This is a test that was frequently used prior to the development of the synthetic test methods described below and consequently, today, this approach should only be used when all other testing alternatives are not suitable.

8.1.6 Synthetic Tests

Synthetic tests [2], are essentially a two part test that is done all at once. The test is performed by combining a moderate voltage source which supplies the full primary short circuit current with a second, high voltage, low current, power source which injects a high frequency, high voltage, pulse at a precise time near the natural current zero of the primary high current.

Effectively, what has been accomplished is to reproduce the conditions that closely simulate those that prevail in the interrupter during the high current arcing and the high voltage recovery periods. As long as the sources, voltage and current, are not appreciably modified, or distorted by the arc voltage then, the energy input into the interrupter, during the high current arcing time region is no different than the energy input obtained from a full rated system current, and voltage, because as we well know, the energy input to the interrupter is only a function of the arc voltage and not of the system voltage.

The behavior of the interrupter in the two classical regions of interest, namely the thermal and the dielectric regions, are evaluated by the high voltage that is superimposed by the injected voltage/current which when properly timed embraces the transition point where the peak of the extinction voltage just appears and the point where the peak of the recovery voltage is reached thus covering the required thermal and dielectric recovery regions.

In general synthetic tests are performed on a single phase basis, and even though schemes have been developed that enable the tests to be made on a three phase circuit, it is only the largest laboratories that are capable of doing so. In the majority of the facilities the high voltage source for these tests is only available on a single phase basis; because in most cases some of the same power limitations that existed for three phase direct test still exist.

As it has been said before the synthetic test method utilizes two independent sources, one, a current source, which provides the high current, and which for all practical purposes is the same source that is normally used for direct tests, and a second, a voltage source, which in most cases consists of a capacitor bank that is charged to a certain high voltage that is dependent upon the rating of the circuit breaker that is being tested.

There are a number of synthetic test schemes that have been developed, but in reality they all are only a variation of the basic voltage, or current injection schemes. In actual practice, what is used most often by all testing laboratories is the parallel current injection technique.

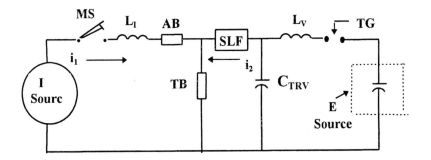

Figure 8.10 Schematic diagram of a parallel current injection synthetic test circuit.

8.1.6.1 Current Injection Method

The current injection method is illustrated in the schematic diagram of the equivalent circuit as shown in figure 8.10. This method is characterized by the injection of a pulse of current that is supplied by the high voltage source.

The high current source, as mentioned before, is composed of a short circuit generator, a back-up circuit breaker for the protection of the test generator, a set of current limiting reactors, a high speed making switch and an additional component, an isolation circuit breaker (IB) whose purpose, as its name implies, is to effectively separate, or isolate, the current circuit from the high voltage circuit.

The high voltage section of the circuit is made-up of a high voltage source (VS) consisting of a capacitor bank that is charged to a predetermined high voltage level. Connected in series with the capacitor is one side of a triggered spark gap (TG) the other side of the trigger gap is connected to a group of frequency tuning reactors. Connected in series with these reactors there is a short line fault (SLF) TRV shaping network, which consist of a combination of capacitors and reactors that in most instances are connected in a classical pi (π) circuit configuration. Generally it is required that at least five of these sections be connected in series in order to accurately represent the TRV of a short line fault. This SLF network however, is only used when the tests that are being performed simulate a short line fault condition.

It is recommended that the frequency for the injected current be kept within the range of 300 to 1000 Hz. These limits depend primarily on the characteristics of the arc voltage. What is important is that the period of the injected current be at least four times longer than the transition period where a significant change in the arc voltage is observed. The magnitude of the injected current should be adjusted so that the rate of change of the injected current $(di/dt)_i$ and the rate of change of the corresponding rated power frequency current

$(di/dt)_p$ are equal at their respective current zeroes. The timing for the initiation of the current pulse is controlled so that the time during which the arc is fed only by the injected current is not more than one quarter of the period of the injected frequency.

Parallel Current Injection. The schematic diagram of the circuit that was shown in figure 8.10 represents the equivalent circuit configuration that is used for the parallel injection method, and in figures 8.11 and 8.12, the relationship between the power frequency and the injected current is shown.

The test is initiated by closing the making switch (MS), which initiates the flow of the current i_1, from the high current source (CS) through the isolating breaker (IB) and the test breaker (TB). As the current approaches its zero crossing the spark gap is triggered and at time t_1, (see figure 8.12), the injected current i_2 begins to flow. The current $i_1 + i_2$ flows through the test breaker until the time t_2 is reached. This is the time when the main current i_1 goes to zero and when the isolation breaker, separates the two power sources. At time t_3 the injected current is interrupted and the high voltage supplied by the high voltage source provides the desired TRV which subsequently appears across the terminals of the circuit breaker that is being tested.

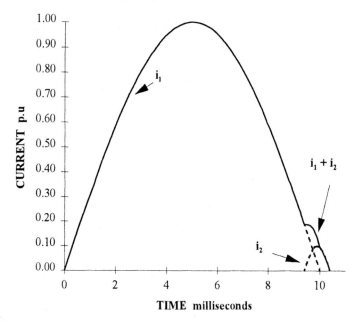

Figure 8.11 Relationship between primary current and injected current in a synthetic test parallel current injection scheme.

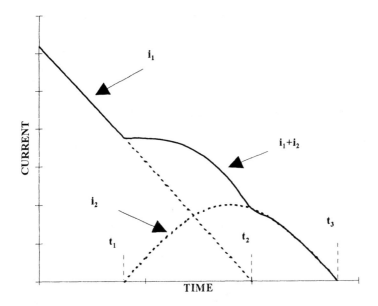

Figure 8.12 Expanded view of the parallel current injection near current zero.

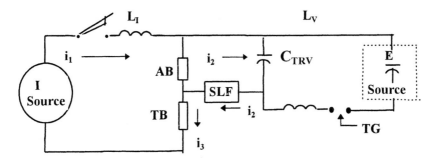

Figure 8.13 Schematic diagram of a typical series current injection synthetic test circuit.

Series Current Injection. The series current injection circuit is shown schematically in figure 8.13 while in figures 8.14 and 8.15 the algebraic summation of the injected currents is shown. The notable difference between the series current injection and the parallel injection methods is that the high voltage source, for the series injection version of the test, is connected is series with the high current source voltage.

At the initiation of the test the making switch is closed and at time t_1 the spark gap is triggered thus, allowing the current i_2 to flow through the isolation breaker but in the opposite direction to that of the current i_1 from the high current source. At time t_2, when the currents i_1 and i_2 are equal and opposite, the current in the isolating breaker is interrupted and during the time interval from t_2 to t_3 the current that is flowing through the test breaker is equal to i_3. This current corresponds to the summation of the currents $i_1 + i_2$ that is produced by the series combination of the high current and the high voltage sources. Following the interruption of the current i_3 at time t_3 the resulting TRV supplied by the high voltage source appears across the breaker terminals.

8.1.6.2 Voltage Injection Method

The voltage injection method, in principle, is the same as the parallel current injection. The only difference is that the output of the high voltage source is injected across the open contacts of the test breaker following the interruption of the short circuit current which, as explained before, is supplied by the high current source. The high voltage is injected immediately after the current zero and near the peak of the recovery voltage that is produced by the power frequency current source. A capacitor is connected, in parallel across the contacts of the isolation breaker, in order to effectively apply the recovery voltage of the current source to the test breaker. This test method is not very popular because it requires a very accurate timing for the voltage injection. This timing becomes a critical parameter which in most cases is rather difficult to control.

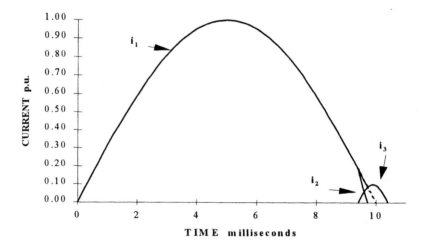

Figure 8.14 Relationship between primary current and injected current in a synthetic test series current injection scheme.

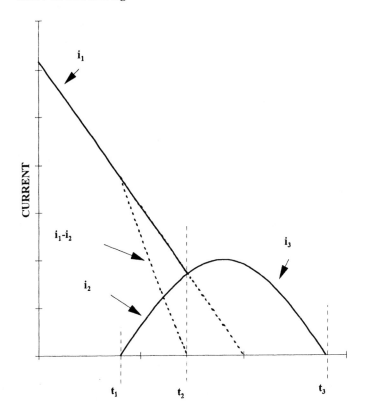

Figure 8.15 Expanded view of the series current injection near current zero.

8.1.6.3. Advantages and Disadvantages of Synthetic Tests

As is the case with any of the other test methods, there are a number of advantages and disadvantages that are associated with synthetic tests. The principal advantage that should be mentioned is that these tests are of a non-destructive nature and therefore they are ideal for development test purposes, where the ultimate limits of the device can be explored without destroying the test model. Also the synthetic test method is the most adequate, and in some cases the only way of performing short line fault tests.

The main disadvantage of synthetic tests is that these tests are primarily a single loop test, which explains why they are considered to be a non-destructive test, and although a reignition circuit can be used to force a longer arcing time, or a second loop of current, it still is very difficult to do a fast re-closing with extended arcing times. Another disadvantage is that this method is not suitable for testing interrupters which have an impedance connected in parallel with the interrupter contacts in which case it is likely that the full recovery voltage can not be attained due to the power limitations of the high voltage source.

8.2 Test Measurements and Procedures

Test procedures and instrumentation, naturally vary in accordance, not only with the test method, but also with the purpose of the tests that are being performed. The test instrumentation, can be significantly different, for example, when doing interrupter development tests, than when doing verification or circuit breaker performance tests.

In the investigative portion of the tests it is likely that special attention will be paid to the phenomena occurring at, or very near, current zero, where a higher degree of resolution is needed. In these tests, what is of interest is what takes place around current zero in a time region which is normally in the microsecond range, while for verification tests, or complete circuit breaker interruption tests, and with the exception of the TRV, the time of interest is in the millisecond range.

For development tests in most cases it is important to have accurate measurements of arc voltage, interrupter pressure, post arc current and other very definite measurements, depending solely on the type of information that is being sought. While, because the tests that are made to demonstrate the capability of the circuit breaker, and the requirements that have been set to demonstrate its compliance with the existing standards, are well defined in the applicable standards [3], [4] and because in most cases it is assumed that a significant number of development tests have already been performed the needed instrumentation is what may be considered as conventional, consisting of measurements of phase currents, phase voltages and TRV.

Since the procedures for development tests are rather specialized and specific in nature according to the circumstances, or to the objectives of the tests being performed it is then difficult to provide firm guidelines for the instrumentation to be used and for the tests procedures to be followed; however, the techniques that are to be described, and which are used for the design verification tests can also be used for other purposes, such as exploratory tests.

8.2.1 Measured Parameters and Test Set-Up

It goes without saying that the fundamental parameters of phase currents and corresponding phase voltages must be measured. In addition to these parameters it is advisable, specially when testing vacuum circuit breakers, to make measurements of the amplified arc voltage. This measurement can be very helpful in determining the precise instant where contact part occurs. It also serves to determine the stability of the arc and the effectiveness of the interrupter at the transition point of the current regions.

Another valuable and important measurement, that sometimes is neglected, is the measurement of the breaker contact travel which, when everything goes right on a test may not be needed but, for those times when there is a failure this measurement would help to answer questions such as: Was the circuit breaker fully open?, did it stall?, was it fully closed?.

The test current is generally measured using a low resistance shunt, and in some occasions, for even better accuracy and response, a coaxial shunt is used. The voltage measurements are usually made with a capacitive compensated voltage divider, and the measurement is preferably made on a differential mode to avoid distortions due to possible ground shifts.

8.2.1.1 Grounding

The one essential requirement is that grounding of the circuit should be either at the source, or at the test breaker but not at both places.

8.2.1.2 Control Voltage

Although in the standards it is indicated that the rated control voltage of the device should be used, one could take exception to this because in most cases it is very convenient to use a higher control voltage, perhaps as much as 20% over the rated value, as the means for minimizing the variation in the time that it takes to open the contacts. In addition to the higher control voltage it is advisable to use a dc supply whenever possible, rather than an ac supply. This is done with the sole objective of minimizing the variation of the contact opening time, which is important because of the need for proper control of the point on the wave where it is desired to break the circuit so that the proper current symmetry is achieved during the test..

8.2.1.3 Close-Open Operations

In some laboratories, performing a close-open operation at high asymmetrical currents can be quite harmful to the health of the circuit breaker, this happens because when, both, closing against the peak of a fully asymmetrical current and opening at the point of maximum total rms. current is being attempted there is a risk that the peak of the closing current may be substantially higher than what is required. This risk exists because, even though super excitation is applied to the test generator, the asymmetrical value at contact part can not be easily achieved and when the symmetrical value of the current is raised to compensate for the rapid dc decrement, the initial peak of the current is also proportionally increased. One method that has been used successfully to overcome this difficulty has been the addition of a series reactor which limits the peak of the current during the closing operation, but as soon as the circuit breaker is closed, and before the opening is initiated, a switch that is connected in parallel with the reactor is closed, effectively shorting out the reactance and thus increasing the current at the time of contact separation.

8.2.1.4 Measuring the TRV

Despite the fact that in most cases the TRV is measured during the actual interruption testing this measurement is not always a valid one. Because of the influences exerted by the characteristics of the arc voltage, the post arc conductivity, and the presence of TRV modifying components such as capacitors and resistors that may have been installed across the contacts and which will most definitely affect the TRV wave of the circuit. Therefore unless the above mentioned effects are insignificant and the short circuit current does not have a dc. component, commonly obtained test records can not be used and special procedures must be utilized to determine the inherent TRV of the test circuit.

The two methods most commonly used are:

1. Current injection
2. Capacitor injection
3. Network modeling/calculation

Current injection. This method consists of injecting a small power frequency current signal into a de-energized circuit and then interrupting the injected current using a switching device that has negligible arc voltage, and post arc current. A device with such characteristics could be a fast switching diode, one that exhibits a reverse recovery time of less than 100 nanoseconds. When using these type of diodes, it is permissible to have a shorting switch across the diode if there is a possibility that the current carrying capabilities of the diode could be exceeded. The shorting switch will have to be opened shortly before the zero current crossing where the TRV measurement is to be made. The measurements of the current and voltage waveforms must be made using instrumentation suitable for high speed recording.

Capacitor injection. Here a low energy capacitive discharge is used as the source for the injected current. In reality this method is no different than the previous method except that in this case the ac. source for the injected current has been replaced by the dc. voltage stored in a charged capacitor. Since the frequency of the discharged current is proportional to capacitance of the source and the inductance of the circuit, then the frequency of the measured voltage defines the inherent TRV.

For best results the frequency of the discharge current for these measurements should be (0.125 of the equivalent natural frequency of the circuit being measured.

Network modeling/calculation. This method consists of either an analog or a digital modeling of the of the network that is being evaluated. The accuracy of this method, of course, depends upon the selection of the appropriate representative parameters of the circuit that is being evaluated.

8.2.2 Test Sequences

As it was the case with regard to ratings, so it is for testing; the ANSI and IEC test requirements are not exactly the same. Nevertheless, the required tests are sufficiently close in both documents, and with only a little extra effort in choosing equivalent test parameters, specially for TRV, and by adding a few extra tests, the requirements of both standards can be concurrently met.

8.2.2.1 IEC 56 Requirements

The short circuit capability, according to the IEC standards, is demonstrated by a test series consisting of five test duties. Test duties 1, 2 and 3 consist of three opening operations which can be demonstrated using the standard duty cycle which, as it can be recalled, consists of the following sequence, O-t-CO-t'-CO where t is either 3 minutes or 0.3 seconds depending on weather the circuit breaker is rated for reclosing duty or not. These test duties are performed with symmetrical currents of 10% ((20%), 30% ((20%) and 60 % ((10%) of the rated short circuit current respectively. The TRV requirements include a slightly higher voltage peak and a significantly shorter time duration to reach the voltage peak. These test duties are performed with the intent of simulating the interrupting behavior of a circuit breaker in the event of a fault in the secondary side of a transformer, where, as the current is reduced the TRV becomes more severe, as it has been verified by Harner et, al. [5].

Test duty 4 consists of the prescribed operating duty cycle. The opening operation is made under symmetrical current conditions, while the maximum asymmetrical current peak must be attained during the closing operation in order to demonstrate the close and latch capability of the circuit breaker. The symmetrical current for the opening following the closing is obtained by delaying the trip sufficiently so that the dc and ac transient components have decayed to an asymmetrical value of less than 20%.

Test duty 5 is a test similar to test duty 4 except that both the opening and closing operations are made with an asymmetrical current. The asymmetry of the current is that which corresponds to a time constant of approximately 45 ms and which corresponds to an X/R value of 14 for 50 Hz. or 17 for 60 Hz.

The asymmetrical value of the current is determined using the actual contact opening time of the circuit breaker to establish the elapsed time that is measured from the time of current initiation to the point of contact separation this time thus, determines the asymmetrical value for the test by following the procedure which was described earlier in chapter 7.

TABLE 8.1
Test for Demonstrating the Short circuit Rating
of a High Voltage Circuit Breaker
(Three Phase Test)

Test Duty	Operating Duty	Phases	V Initial & Recov.	Making Current at 1st. Major Loop		Current Interrupted at Contact Part	
				A, rms	A Peak	A rms	% Asym
1	One O & One CO	3	V			.07 to .13 I	> 50
2	One O & One CO	3	V			.2 to .3 I	< 50
3	One O & One CO	3	V			.4 to .6 I	> 50
4	O-15s-O, O-15s-CO or CO-15s-CO	3	V			I	< 50
5	O-15s-O, O-15s-CO or CO-15s-CO	3	V/K			KI	< 20
6-1	CO-15s-CO	3	V	1.6 I	2.7 I	SI	> 50
6-2	C	3	V	1.6 I	2.7 I		
6-3	O-15s-O	3	V			SI	> 50
	121 kV and above						
7A-1	CO-15s-CO-15m-CO-1h-CO	3	V/K	1.6 KI	2.7 KI	KSI	> 50
7A-2	C-15s-C-15m-C-1h-C	3	V/K	1.6 KI	2.7 KI		
7A-3	O-15s-O-15m-O-1h-O	3	V/K			KSI	> 50
	All other breakers						
7B-1	CO-15s-CO-1h-CO	3	V/K	1.6 KI	2.7 KI	KSI	> 50
7B-2	C-15s-C-1h-C	3	V/K	1.6 KI	2.7 KI		
7B-3	O-15s-O-1h-O	3	V/K			KSI	> 50
8	Several O and CO	3	V/K				Random
9	O-0s-CO or CO-0s-CO	3	V			RSI	> 50
10	O-0s-CO or CO-0s-CO	3	V/K			RKSI	> 50
11	C-T s-O	3	V/K	1.6 KI	2.7 KI	KI	
12	In closed Position	1					
13	1-O and 1 CO or 2 O	1	.58V			Smaller of 1.15 I or KI	< 20
14	1-O and 1 CO or 2 O	1	.58V			Smaller of 1.15 I or KSI	> 50
15	O-15s-O or O-15s-CO	1	.58V			.7 to .8 I	< 20
16	O-15s-O or O-15s-CO	1	.58V/K			.9 to .95 I	< 20

8.2.2.2 ANSI C37.09 Test Sequences

The ANSI test requirements are specified in Tables 1 and 2 of the C37.09 test standard for high voltage circuit breakers. The included Table 8.1, corresponds to Table 1 of the above mentioned test standard, except that the last two columns have been omitted. As it can be seen from Table 8.1, a rather extensive test series would be required if every test is performed precisely as described; however, as it will be shown later, when one looks closely to the requirements it is found that some of the tests can be combined, while for others, alternate methods are used due to limitations of the testing facilities.

The first three ANSI sequences are quite similar, if not identical to those required by IEC 56. They are composed of one opening and one close-open operation at reduced current values. The ANSI current values are specified within a percentage range that encompasses the specific percentage values given by IEC 56, however, for test duties 1 and 3 ANSI specifies asymmetrical currents. But as it is presently recognized and agreed upon; that, given the intent of these tests, which as it was previously stated, are meant to simulate a secondary fault and to demonstrate the breaker capability to withstand higher rates of TRV, the tests should be performed with symmetrical currents.

Test duties 4 and 5 are symmetrical current tests, both are basically the same in terms of operating sequence, except that test 5 is performed at the lower operating voltage and the higher short circuit current as defined by the rated voltage range factor K. This test, consequently, is only applicable to indoor circuit breakers since the K factor for all other breakers is equal to 1. Furthermore, as it has been repeatedly said, for those circuit breakers that utilize modern technologies, the inverse relationship between voltage and current is no longer applicable, and as it is usually the case, the maximum current can be interrupted at the maximum voltage. What this implies is that, in most cases, for newer technology circuit breakers it may be possible to omit test duty 5, provided that the higher current is used. Figure 8.16 shows a typical oscillogram of this particular test duty

It is important to note that some alternatives are allowed for this test cycle, one being a Close-Open—15 sec—Close-Open sequence however, it is advisable to avoid using this sequence simply because it is a little more difficult to control the symmetry of the current during the opening operation that follows immediately after the closing operation, this is so even though, the symmetry requirements are applicable only to one opening and one closing operation. Since it is easier to adjust the time of the opening of the contacts for an initial open only operation, one should take advantage of this condition and subsequently control the closing of the contacts so the desired value of asymmetry is achieved leaving the remaining opening test duty operation to occur after a short time delay needed for the current to regain its symmetry.

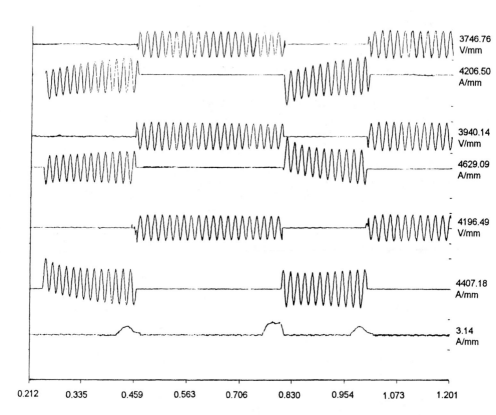

Figure 8.16 Typical record oscillogram of a three phase Close-Open—0.3 sec.—Close-Open operation with an asymmetrical current.

The above comments are applicable not only to this particular test duty but also to all of those tests that start with a closing of the test circuit breaker, or that require multiple operations. Furthermore, it is also possible, as it is done in IEC 56, to perform this test duty with a time interval of 0.3 seconds and thus meeting the test requirements for a fast reclosing duty by essentially having combined test duties 4 and 9. However, when this done the possible need of derating the interrupter must be considered. Derating was almost always applied to older style circuit breakers, but when considering the application of the new technologies, this is a requirement that does not appear to have much value left.

Test duties 6, 7A and 7B, are also a demonstration of the standard test duty cycle, except that in test duty 7A, which is intended for circuit breakers rated above 121 kV, a second test duty cycle is performed after 15 minutes of the first. Remembering that these standards were written, primarily with air and oil circuit breakers in mind, it is understandable that the possible effects of the stored heat and the drop in pressure due to the previous interruption had to be investigated. With today's vacuum or SF_6 circuit breakers it is known that this does not constitute a problem and therefore the only justification for the extended duty cycles is to accumulate the required 800% interrupted currents, and the demonstration of the worst switching condition. The time interval between duty cycles also is no longer that important and the tests can be made within a time frame of only a few minutes as it may be dictated strictly by convenience. Again, when performing these test duties it is recommended to start with an opening rather than with a close-open operation, and that is so the asymmetry of the current can be properly controlled. It is also important to note that the high asymmetrical values, those above 50%, that are specified in the test tables are unrealistically high, for a circuit breaker with normal interrupting times of 3 or 5 cycles and for a system having an X/R value of 17, and therefore the alternative is to test with a lower asymmetry, say 35 to 45%, or as is suggested by the standards, to adjust for the total current at the time of contact part but to reduce the symmetrical rms. of the current and test with the higher total rms. asymmetrical current value. Nevertheless, a test made with a reduced symmetrical rms. and with a higher dc component is valuable because it provides at least some indication of the breaker capability for applications in systems where the X/R values are higher than 17.

Test duty 11 calls for the circuit breaker to be closed against a full value of a fault current, to carry this current for a time equal to the maximum rated permissible trip delay (2 seconds for breakers rated 72.5 kV and below, or 1 second for higher voltage ratings) and then to interrupt the full rated symmetrical short circuit current. This is a very difficult test to perform, mainly, because the thermal rating of the test generators are almost always exceeded, and this kind of power is simply not available on a three phase basis at any test laboratory, for the higher voltage rated circuit breakers.

In some laboratories this test may be performed synthetically, using two power sources, the closing operation is done using a high voltage and high current source; this initial closing portion of the test is obviously, no different than any of the routine closing operations that are performed as part of the other required test duties. However, immediately after the circuit breaker is closed the high voltage source is removed by means of an isolating switch and then a high current, that is supplied by a low voltage source, is superimposed upon the closed circuit breaker contacts. The high current is maintained for the required length of time and afterwards, when the time requirements are met, the high current source is removed and the high voltage high current source is once again inserted into the circuit so that the current interruption portion may be performed.

Prior to the publication of the 1964 edition of the standards C37.04 and C37.09 the requirements for test duty 11 were promulgated separately, and a close and latch, and a momentary current rating were published. The testing to demonstrate these requirements was done independently and in separate operations. This approach is still taken in many cases because of a tacit agreement about the impracticality of the present requirements. What it must be remembered is that the close and latch test is made to prove the mechanical capability of the circuit breaker and that this demonstration is made several times while performing the complete test sequences required for breaker certification. The requirement for carrying the current for an specific time duration (longer than that required in the previous test duty) is met in the very next test sequence, where the test demonstrates the short time thermal capability of the contacts. It is generally agreed that if this capability is built-in into the contact structure, the higher contact temperature at the moment of contact part does not have any negative effect upon the interrupting capability of the circuit breaker.

Test duties 13 and 14 are supposed to demonstrate the capability of the circuit breaker to interrupt a line to ground fault in a grounded system. The tests, naturally, are performed on a single phase basis and are only required when the previous test duties have been done on a three basis, these tests are not required when all the testing is done using a single phase source.

What it is important to realize is that these two tests are rather unique because they are performed at a voltage of .58% of the maximum line to line rated voltage and that this is in contrast with the .87% of the line to line voltage which is used when all the testing is done with a single phase source. Also the specified fault current, for test duties 13 and 14, was chosen by applying the inverse voltage-current relationship and where the factor 1.15 is simply the ratio between the maximum voltage V and 0.87, $(1/0.87 = 1.15)$; however. this same criteria is not applied to the test series when made on a single phase basis.

Finally, it is not clear what the TRV value should be for test duties 13 and 14, however what is generally agreed is that the peak of the recovery voltage should be between 1.17 and 1.20 times the maximum rated line to line voltage.

It should be expected that in the near future, as the ANSI standards are revised and as the harmonization process with the international standards progresses further, these small inconsistencies will probably disappear.

The last two test duties, 15 and 16, are also single phase tests, but are applicable only to outdoor circuit breakers. The aim of these tests is to prove the short line fault capability of the circuit breaker. However, in regards to these test duties, another inconsistency is found between two of the ANSI documents. C37.04 states that all outdoor circuit breakers must be capable of interrupting a short line fault, but in C37.09 it is said that it is not needed to demonstrate this capability for circuit breakers rated 72.5 kV and below.

Experience has shown that the short line fault requirements are not confined only to the very high voltage circuit breakers and, as a matter of fact, a number of circuit breaker failures which can be directly attributable to the inability to properly handle the short line TRV have been reported. This fact is now widely recognized and even though, it is still unofficial, short line fault testing is being performed in all outdoor circuit breakers, regardless of their voltage rating.

8.2.2.3 Most severe switching conditions

The most severe switching conditions are generally referred as to the case where the interrupter is subjected to a maximum arcing time, or what it amounts to a condition of maximum arc energy input. Basically what it is intended by testing for the most severe switching conditions is to show that in the worst case the interrupter in any one of the poles of the circuit breaker is capable to withstand the maximum arc energy input.

The most unfavorable conditions will be those where the contact separation occurs during a minor current loop and where the duration of the arcing time is just short of the minimum arcing time required for interruption by that particular design. As we already know, if the minimum arcing time requirement is not met then interruption will only take place after an additional half cycle of current, which for the worst case condition will constitute a major current loop.

Since in a three phase system under symmetrical current conditions the current zeroes occur at a sixty electrical degrees interval then, there is a 2.77 millisecond window to accommodate the variation in the possible arcing time. What this means is that, with symmetrical currents in a three phase system, if one of the phases fails to clear the fault at its first current zero, this phase most likely will never see its true maximum arcing time because one of the other phases is likely to interrupt the current before the original phase reaches a repeat current

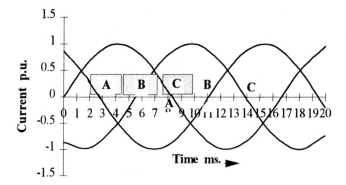

Figure 8.17 Relation of arcing time for symmetrical currents and for different contact parting windows for a circuit breaker with a minimum arcing time of 4 ms.

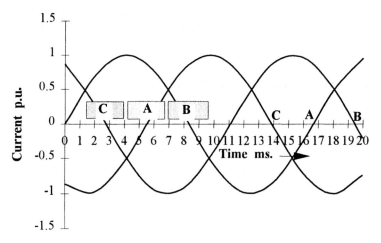

Figure 8.18 Relation of arcing time for symmetrical currents and for different contact parting time windows for a circuit breaker with a minimum arcing time of 10 ms.

zero and, even though, the energy input to the interrupter which failed to clear the current at its first attempt continues to increase, because when the current in one of the phases has been interrupted the remaining two phases will evolve into a single phase current which is then interrupted by the two remaining poles in series, the total energy input still will be less than what can be expected from a fully asymmetrical single phase fault that has a maximum arcing time.

rent zero where interruption should occur is designated by the letters A, B, and C. This designation matches the identification that is given to the corresponding arcing windows. Figure 8.17 represents a circuit breaker that has a minimum nominal arcing time of approximately 4 ms. This arcing time is generally a characteristic of vacuum interrupters with currents greater than 15 kA.

Figure 8.18 shows a minimum arcing time of 10 ms. which is representative of a SF_6 circuit breaker, where the range of the minimum arcing times is generally between 7 to 13 ms.

One suggested method that can be used to determine the maximum arcing time is illustrated in figure 8.19. For the first interruption the contact part is adjusted so that it occurs at a current zero of any of the three phases, in our example phase A was selected first. By observing figure 8.19 we can see that: a) if the minimum arcing time is less than 3 ms. then interruption will take place at B. b) if the minimum arcing time is less than 6 ms. then interruption will occurs at C, and c) if the minimum arcing time is less than 8 ms. then the current will be interrupted at A. For the second test the contact part is advanced by approximately 2.5 ms. to t_2 and in this way the arcing time window, that we had mentioned before is not exceeded and if the test is repeated the point of interruption will be the same as in the previous test, any further advances of the contact part will then result in a shifting of the corresponding current zero where interruption takes place.

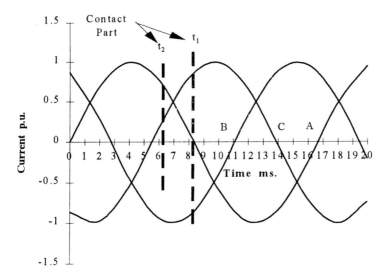

Figure 8.19 Method for obtaining the maximum arcing time for a symmetrical three phase current test.

The same situation, as it was described for the symmetrical currents, does exist with asymmetrical currents, except that now the arcing time window is no longer a constant 2.77 ms. but it depends upon the dc. and ac. components of each of the phase currents. Figures 8.20 and 8.21 serve to illustrate the shift of the interruption point for a circuit breaker that has a minimum arcing time of 4 ms., when the contact parting point is displaced by about 5 ms. What it should be noticed is that for the conditions shown in the figures the maximum energy input to the interrupter, shown, in arbitrary per unit values; does not occur on the first phase to clear but rather on one of the last phases to interrupt. However, there are now two phases in series that are interrupting the current thus making the interruption an easier task.

Of the procedures given, by the respective testing standards, for obtaining the worst switching condition, the one described in IEC 56 provides a more clear and definite approach and it yields the same results that are being sought by ANSI.

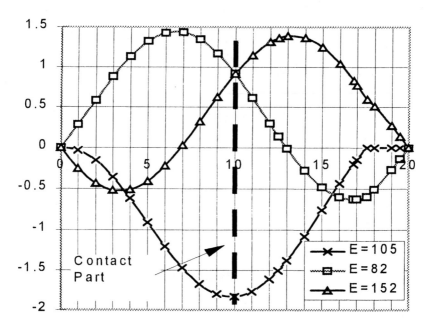

Figure 8.20 Arcing time variation depending on point of contact part for asymmetrical currents. Assumed minimum arcing time 4 milliseconds, a comparative value of arc energy input E (arbitrary per unit value) is shown in enclosed box.

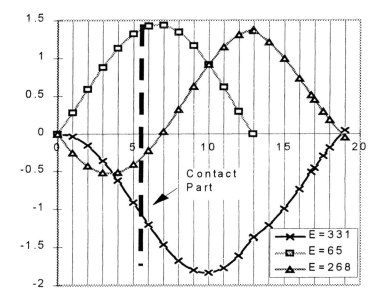

Figure 8.21 Arcing time variation produced by advancing approximately 4.5 ms. the point of contact part from the original position shown in figure 8.21. Assumed minimum arcing time is still 4 milliseconds, and the same comparative value of arc energy input E (arbitrary per unit value) is shown in enclosed box.

The referenced test procedure calls for the following sequences:

1. for the first operation the point of contact part is set so that the required value of the total current is obtained
2. for the second test, the initiation of the short circuit current is shifted by 60 electrical degrees and if in the first test the first phase to clear did so after a major current loop, the trip time is advanced by approximately 130 electrical degrees otherwise it is advanced by only 25 degrees
3. for the third operation the procedure of the second operation may be repeated and the same criteria about the first phase to clear is applicable.

The only objection that perhaps may be raised about the procedure is in connection with the change that is required of the inception angle of the short circuit current, which is required so that the asymmetries of the currents are transposed between the phases, it seems that it would be simpler to change only one parameter, the contact parting time, rather than two parameters at the time, considering that the results for similar contact part conditions for each individual phase, would be the same.

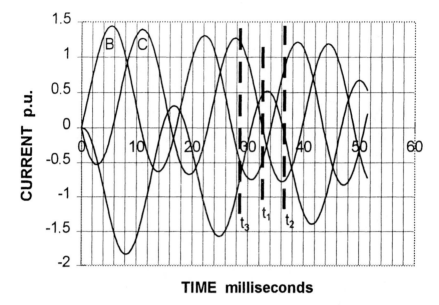

TIME milliseconds

Figure 8.22 Three phase asymmetrical currents times t_1, t_2, and t_3 show the changes in the contact parting time to obtain the required maximum energy conditions Contact parting times are advanced, or delayed by approximately 4.2 ms (45 electric degrees) for a 3 cycle breaker.

In figure 8.22 it is shown how one may accomplish the required interruption of the current after a portion of a minor loop plus a full major loop, by only varying the point of contact part. This is done by controlling the tripping of the circuit breaker so that, for our example which corresponds to a circuit breaker with a minimum arcing time of 4 ms., the contacts will separate on the phase with the highest asymmetry (phase A in our example), at a point on the minor loop that is less than 4 ms. from its next current zero, and which in the case that is illustrated corresponds to a current that is close to the peak of the minor loop; interruption then will take place after the full major loop. For the next test the trip is advanced by 4.2 ms. and the interruption will occur on phase C, and finally for the last test the trip signal is retarded by 4.2 ms. and interruption then will occurs in phase B. For other circuit breakers having longer arcing times a similar procedure can be used to determine the required changes in the contact parting time.

Another point that one may find to be questionable, and that is only because of the performance characteristics of today's technologies, is the need for satisfying the maximum arcing time requirements while performing test duties 1, 2 and 3.

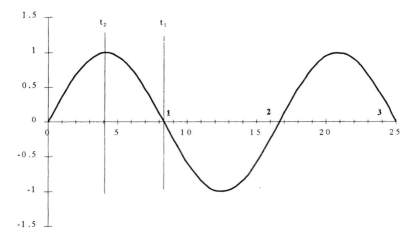

Figure 8.23 Graphical representation of a method for obtaining maximum arcing times during single phase symmetrical current tests.

The currents, and the energy input during these test duties is relatively low and therefore it seems that a better option could be to do these tests with the minimum arcing time rather than with the maximum arcing time, since at these lower current levels it may be expected that the interrupter may quench the arc sooner and with an shorter gap which may not be able to withstand the full recovery voltage.

A more realistic and an easier demonstration for the maximum arcing time capabilities of an interrupter is obtained with a single phase test because, in a single phase test the arcing time can be controlled fairly without having to be concerned about the interference from the other phases in the event that the interrupter fails to interrupt at the first current zero.

For single phase tests with symmetrical currents (refer to figure 8.23), the maximum arcing time can be obtained by first, adjusting the contact part to coincide with a current zero crossing (at time t_1), interruption will occur at either the first current zero (point **2**), for circuit breakers that have minimum arcing times of less than eight milliseconds, or at the second current zero (point **3**) for circuit breakers with arcing times greater than eight milliseconds.

For the next test, the point of contact part (t_2) is advanced by approximately 4.5 milliseconds, under these conditions interruption will take place at the first current zero (point **1**) for breakers with minimum arcing times of less than 4 milliseconds (vacuum circuit breakers for example), at the second current zero (point **2**) for circuit breakers that have a minimum arcing time greater than 4 milliseconds but less than 12 milliseconds, or at the third current zero (point **3**) for circuit breakers that have minimum arcing times greater than 12 millisec-

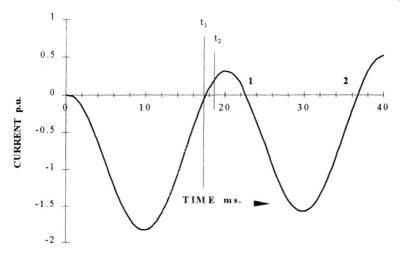

Figure 8.24 Graphical representation of a method for obtaining maximum arcing times during single phase asymmetrical current tests.

onds. Applying the above described procedure will ensure that the interrupter has been subjected to the longest possible arcing time when interrupting symmetrical currents.

For an asymmetrical condition, see figure 8.24, a similar procedure may be used as follows:

First, the region where the contacts must part to satisfy the total current requirements should be chosen, this time is then adjusted to coincide with the beginning of the minor loop shown as time t_1, if interruption occurs at the first current zero (point **1**) the time should be retarded by about 3 milliseconds, to time t_2, interruption then will most likely take place at current zero corresponding with point **2**. This last test will demonstrate the maximum arcing time conditions for circuit breakers that have minimum arcing times in the range of 4 to about 12 milliseconds. Since this is generally the range of minimum arcing time of modern circuit breakers the above test will be sufficient to verify the maximum energy input condition within reasonable limits of accuracy for most of today's vacuum and SF_6 high voltage circuit breakers.

REFERENCES

1. Walter W. Wilson, Casjen F. Harders, Improved Reliability From Statistical Redundancy of Three Phase Operation of High Voltage Circuit Break-

ers, IEEE Transactions of Power Apparatus and Systems, Vol. PAS-90, No. 2: 670-681, March/April 1971.

2. ANSI/IEEE C37.081-1981, IEEE Application Guide for Synthetic Fault Testing of AC High-Voltage Circuit Breakers Rated on a Symmetrical Current Basis.

3. ANSI/IEEE C37.09-1979, IEEE Standard Test Procedure for AC High-Voltage Circuit Breakers Rated on a Symmetrical Current Basis.

4. IEC 56, High-voltage alternating-current circuit-breakers.

5. R. H. Harner, J. Rodriguez, "Transient Recovery Voltages Associated with Power-System Three Phase Transformer Secondary Faults" Transactions of Power Apparatus and Systems : 1887-1896, November/December 1972.

6. IEEE Tutorial Course, Application of Power Circuit Breakers, IEEE Power Engineering Society, 93 EHO 388-9-PWR: 74-84.

7. W. P. Legros, A. M. Genon, M. M. Morant, P. G. Scarpa, R. Planche, C. Guilloux, Computer Aided Design of Synthetic Test Circuits for High Voltage Circuit Breakers, IEEE Trans. Power Deliv. Vol. 4, No. 2: 1049-1055, April 1989.

8. G. St.-Jean, M, Landry, Comparison of Waveshape Quality of Artificial Lines Used for Short Line Breaking Tests on High Voltage Circuit Breakers, IEEE Trans. Power Deliv. Vol. 4, No. 4: 2109-2113, Oct. 1989.

9. L. M. J. Vries, G. C. Damstra, Areignition Installation with Triggered Vacuum Gaps for Synthetic Fault Interruption Testing, IEEE Trans. Power Deliv. Vol. PWRD-1, No. 2: 75-80, April 1980.

9

PRACTICAL CIRCUIT BREAKER APPLICATIONS

9.0 Introduction

In an earlier chapter, a circuit breaker was defined as: "a mechanical device which is capable of making, carrying and breaking currents." It was also learned that there are a number of different types of circuit breakers and a number of different system conditions where they may be applied and, that as a consequence, it is to be expected that unusual conditions, or situations that deviate from what is considered to be an standard, or normal condition can be encountered, and that in most instances these non-standard conditions will have a significant effect upon the application of a circuit breaker.

The primary aim of this chapter is to provide some simple and practical answers to questions relating to non-standard applications, so they can be used to facilitate the evaluation of a given circuit breaker for a given application. Naturally, there are so many unique conditions that it will not be possible to cover all of the foreseeable applications; but, we will concentrate on those that are most frequently encountered.

Among the most fundamental and often asked questions about circuit breaker applications are those relating to:

a)   Overload currents and temperature rise
b)   High X/R systems
c)   Systems with frequencies other than 50 or 60 Hz.
d)   Size of capacitors banks for capacitor switching operations
e)   High TRV applications
f)   High altitude installations
g)   Low current, high inductive load current switching
h)   Choice between SF_6 or vacuum.

9.1 Overload Currents and Temperature Rise

The continuous current carrying rating of a circuit breaker is predicated on the premise that the ambient temperature and the elevation where the circuit breaker is applied is within the limits that have been set by the applicable standards. As ambient temperatures vary widely on a daily and on a seasonal basis, to provide a constant base of reference an ambient temperature of 40°C was selected as the upper limit. This selection was based on the meteorological reports provided by the US Weather Bureau, which indicate that the ambient temperatures in the continental US very seldom exceed this upper limit. The maximum standard altitude as it will be recalled is 1000 meters (3300 feet), over sea level. This elevation is considered to be within the limits of the standard operating conditions because the majority of the applications world wide do not exceed this limit..

The altitude limitations are related to the lower density and therefore lesser cooling capability of the air at higher elevations; while, the ambient temperature is directly related to the total temperature of the equipment which is dictated by the limitations that are established based by the characteristics of the materials that are employed in the construction of the circuit breaker.

To evaluate the behavior of the circuit breaker under conditions which are deemed to be different than those considered as standard; be them larger currents, higher ambient or higher altitudes, the problems reduces to one of establishing the ultimate temperature rise required to dissipate, by convection and radiation losses the watts generated at specific currents.

For electrical equipment that has only few ferrous materials components the losses are essentially proportional to the square of the current. However, as the temperature increases, so does the resistance and if the losses were due to the conductors alone then the loss curve will rise slightly faster than the square function. But in most circuit breakers there is a significant amount of ferrous components and the losses due to eddy currents are approximately proportional to the 1.6 power of the current. Considering these to values to be the extreme limits and based primarily on practical experience an exponent value of 1.8 has been established as a suitable compromise.

When the circuit breaker has reached its ultimate temperature rise for a given steady state current it is clear that the total losses must be dissipated since the equipment is then no longer storing any of the generated heat. These losses are divided essentially into radiation and convection losses. The former varies approximately as the difference, raised to the fourth power, of the absolute temperatures, while the latter varies at a much lower power of the temperature. The above statements about the losses are given only as general reference and it is not implied, nor is it necessary to calculate these dissipation factors before solving the problem at hand.

9.1.1 Effects of Solar Radiation

For outdoor applications, in addition to the heating produced by the load current and by the ambient air temperature, one must be aware of the possible additional heating that may result from the effects of solar radiation. On the basis of field tests and accumulated operating data it has been determined that in most cases a maximum temperature rise of approximately 15°C (27°F) may be expected on the conducting parts of the circuit breaker.

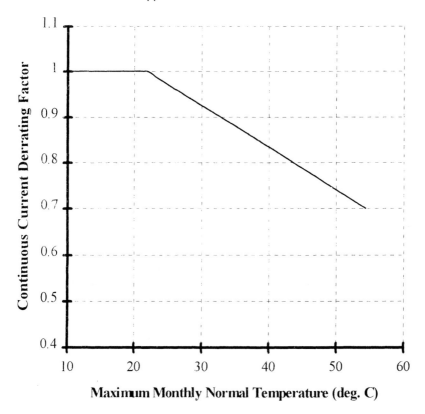

Figure 9.1 Altitude correction factors for continuous currents.

When the circuit breaker is operated at a monthly normal maximum ambient temperature above 25°C (77°F) derating of the continuous current capability of the circuit breaker may be necessary. The derating factor to be used [1] is given in figure 9.1 as a function of the maximum monthly normal temperature as given by the US Weather Bureau.

9.1.2 Continuous Overload Capability

There are times when it becomes necessary to operate a circuit breaker with load currents that are higher than those corresponding to the full rating of the circuit breaker. Operation under these conditions is possible provided that the ambient temperature is consistently below the maximum allowable 40(C.

To find the allowable current that can be carried at a given ambient temperature the following equation is given.

$$I_a = I_r \left(\frac{\theta_{max.} - \theta_a}{\theta_r} \right)^{\frac{1}{1.8}}$$

where

I_a = allowable current
I_r = rated continuous current
θ_{max} = allowable hottest spot total temperature
θ_a = actual ambient temperature
θ_r = allowable hottest spot total temperature rise at rated continuous current

In order to prevent that the maximum temperatures at any given point and for any given material are not exceeded when the load current is adjusted to compensate for the lower ambient temperatures the following rules should be observed [2].

1. If the actual ambient temperature is less than 40°C, the component with the highest specified values of allowable temperature limitations should be used for determining $\theta_{max.}$ and θ_r .

2. If the actual ambient is greater than 40°C, the component with the lowest specified values of allowable temperature limitations should be used for determining $\theta_{max.}$ and θ_r .

By using these values for the calculations it is assured that the temperatures of any parts of the circuit breaker would not be exceeded. However, in many cases it would perfectly safe to exceed these limits without the risk of impairing the performance or the life of the circuit breaker. This is so because generally the minimum limits of temperature are at breaker locations that are readily accessible to the operating personal while the maximum temperatures are allowed at external locations, that are not accessible to operating personnel. If these parts are excluded from consideration, higher values of permissible currents will be obtained from the calculations but, these calculated values should be used judiciously and only when the particularities of the design are well known to insure that there is no possibility of damage to adjacent lower temperature materials.

As an example, let us consider a circuit breaker that has a continuous current rating of 1200 A. This circuit breaker is going to be applied at an ambient temperature of 25°C. The maximum allowable temperature rise is limited to 65°C by its bushings. It is desired to find what is the maximum current capability for this breaker under the given conditions.

Using the given equation, the maximum allowable current for this breaker is determined to be 1246 amperes.

9.1.3 Short Time Overloads

The permissible time duration of overload currents, which are predicated upon a specific temperature ceiling, are intimately concerned with the thermal capacities of the components and hence with the rate of growth of temperature with time. Therefore, to determine what would be a safe overload, in terms of current or time, we must find the interrelation of the above three factors: the watts that are generated at specific currents, the ultimate temperature rise required to dissipate by convection and by radiation losses and the nature of the growth of the temperature with respect to time, towards the ultimate temperature rise for any particular current.

For simple structures, where circuit breakers may be considered to be one, it is fairly accurate to assume that the temperature increases exponentially towards the ultimate temperature rise. This means that the growth of the temperature is progressing in such a way that it is continuously consuming a fixed proportion of the remaining temperature rise in equal intervals of time. The exponential temperature rise curve reaches 63% of its remaining rise in an interval of time equal to its time constant τ.

The time constant on critical circuit breaker parts generally falls between 30 to 90 minutes, and this value may be specified by the circuit breaker manufacturer, but if not, it would be safe to use a value of 30 minutes.

To calculate the time duration of a short-time overload the following equations should be used.

$$t_s = \tau \left[-\ln \left(1 - \frac{\theta_{max} - Y - \theta_a}{Y\left[\left(\dfrac{I_s}{I_i} \right)^{1.8} - 1 \right]} \right) \right]$$

$$Y = \left(\theta_{max} - 40^\circ C \right) \left(\frac{I_i}{I_r} \right)^{1.8}$$

where

θ_{max} = maximum allowable total temperature °C
θ_a = actual ambient temperature °C

I_i = initial current carried by the breaker during the preceding 4 hours
I_s = short time load current in amperes
I_r = rated current in amperes
τ = thermal time constant of the circuit breaker
t_s = permissible time in hours for carrying overload current

The emergency load current capability for a circuit breaker is treated on the referenced application standard [2] by establishing emergency load current carrying capability factors which are based on an ambient temperature of 40(C, for two distinct overload allowable periods; a four hour and an eight hour period, for the numerical values of these factors refer to figure 9.2.

According to the rules, it is permissible to operate 15°C above the limits of total temperature for the four hour period and 10°C for the eight hour period.

The following guidelines are a direct quote from the referenced standard:

Each cycle of operation is separate, and no time-current integration is permissible to increase the number of periods within a given cycle. However, any combination of separate four hours and eight hours emergency periods may be used, but when they total sixteen hours, the circuit breaker shall be inspected and maintained before being subjected to additional emergency cycles.

For ambient temperatures other than the 40°C maximum specified, the procedures that were previously outlined may be used.

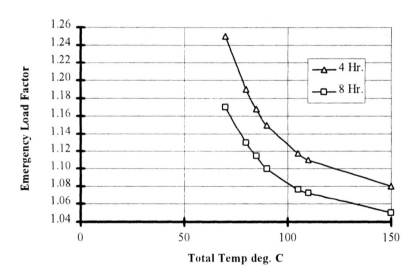

Figure 9.2 Overloading factors for four and eight hours intervals.

Four-Hour Factor. This factor shall be used for a cycle of operation consisting of separate periods of no longer than 4 hours each, with no more than four such occurrences before maintenance.

Eight-Hour Factor. This factor shall be used for a cycle of operation consisting of separate periods of no longer than eight hours each, with no more than two such occurrences before maintenance.

Each cycle of operation is separate, and no time-current integration is permissible to increase the number of periods within a given cycle. However, any combination of separate four hours and eight hours emergency periods may be used, but when they total sixteen hours, the circuit breaker shall be inspected and maintained before being subjected to additional emergency cycles."

For ambient temperatures other than the 40°C maximum specified, the procedures that were previously outlined may be used.

9.1.4 Maximum Continuous Current at High Altitude Applications

Generally, applications at high elevations do not pose much of a problem because the interrupters that are used in today's circuit breakers are sealed devices and consequently the contact structure itself is not affected by the high altitude and the lower air densities. Those parts of the circuit breaker which are exposed to the outside atmosphere are not generally the most critical parts and more importantly as the altitude increases it is less likely that the ambient temperature would reach the 40°C upper limit. In the event that it is desired to calculate the maximum allowable current at high elevations the appropriate multiplying factor that is plotted in figure 9.3 as a function of the maximum monthly normal temperature as given by the US Weather Bureau. As it may be seen in the figure, even at 3,000 meters (10,000 ft) and at ambient temperatures of 35°C the circuit breaker is capable of carrying its full rated continuous current.

To determine the short time overload characteristics of a circuit breaker it is possible to calculate what the overload would be at sea level, and then multiply this value by the factor obtained from figure 9.3 for the corresponding ambient temperature.

9.2 Interruption of Current from High X/R Circuits

As it has been explained before the short circuit ratings assigned by the standards are based on an X/R value of 17 at 60 Hz. or 14 at 50 Hz. This naturally constitutes only a compromise average value, which is representative of the majority of the applications. But that still leaves a significant number of applications where the X/R of the system is greater than the values adopted by the standards. When this happens then these questions arise: What rating do I need in the circuit breaker? What is a circuit breaker I have good for?

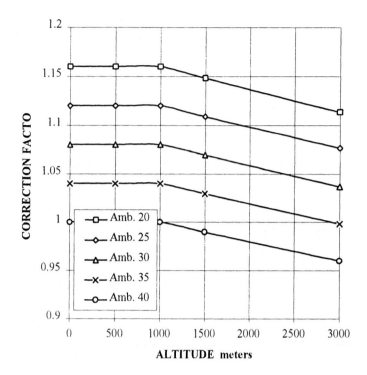

Figure 9.3 Altitude correction factor.

First, it should be remembered that circuit breaker ratings are based on the symmetrical current values and that these symmetrical current ratings are the values that should not be exceeded. However, it is also known that the current asymmetry is a function of the time constant, or X/R of the system and therefore, for a constant contact opening time of a circuit breaker, the total rms. current at the point of contact separation increases as a function of the increase of the asymmetry which in turn is the result of the increase in the time constant of the circuit.

Whenever a fault occurs at a location that is physically close to a large power generator, there may be a significant ac exponential component of the asymmetric current which decays very rapidly during the first few cycles after the initiation of the fault. However this ac decay is generally considered not to be significant at locations that are distant from the power generator, where the short circuit current is fed through two or more transformations, or those applications where the reactance of the system is greater than 1.5 times the subtransient reactance of the generator.

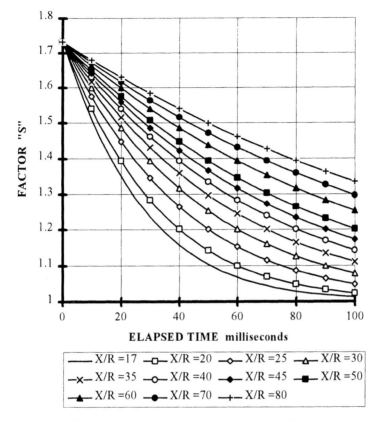

Figure 9.4 Factor "S" for asymmetrical current values with dc decrement only.

For high voltage circuit breakers applications, in practically all cases, it is possible to ignore the effects of the ac transient component and to consider only the dc component. Obviously, this introduces some error in the calculation, especially in the distribution class circuit breakers, where there are perhaps only a few instances where closer attention should be paid to the effects of the ac transient component. The expected errors however, would be in the conservative side and the results would lead to specify a circuit breaker with a higher rating thus assuring a greater margin of safety. We must also recognize that the error is within what can be considered to be acceptable operating limits since a rigorous mathematical analysis of the complete circuit is not feasible and moreover, where the data available for the values of the components very rarely would have an accuracy better than 10%. Furthermore, there is something to be said about operating experience which has shown that this is a relatively conservative and valid approach.

For the discussions that follow, and to answer the two questions posed earlier, a plot of the ratio of the total rms. asymmetrical to the symmetrical rms. current at contact separation, plotted, as a function of the elapsed time, from fault inception for different X/R values is given in figure 9.4. The ratio between the two currents is called the factor "S" and is to be used as a multiplier to establish the related values between the symmetrical and the asymmetrical currents or vice versa.

The first step on the application process is to determine the magnitude of the short circuit current which can be calculated using either of the methods that were given in chapter 2. Next, it is necessary to calculate the X/R value for the circuit.

If the X/R value is equal or less than 17 then it is possible to simply choose a circuit breaker with a symmetrical current interrupting capability that is equal or greater than the calculated short circuit current.

If the X/R factor of the circuit is greater than 17, then it is necessary to determine the elapsed time, or contact parting time, which according to its definition is equal to a one-half cycle relay time plus the contact opening time of the circuit breaker. Once this value has been established, from figure 9.4 the S factors may be determined for the calculated X/R and for the standard value of 17. Multiply the calculated short circuit current by the S factor that corresponds to the higher X/R to obtain the total rms value of the asymmetrical current then, divide this value by the factor S corresponding to the X/R of 17. This is the minimum interrupting current rating that is needed for this particular application.

For example, let us assume a 121 kV circuit capable of delivering a short circuit current of 14,000 amperes and having an X/R of 50. It is desired to select, from a table of preferred ratings, a 5 cycle circuit breaker with a contact parting time, or elapsed time of 50 milliseconds. From figure 9.4 the S value for an X/R of 50 is 1.39 and for X/R of 17 is 1.1.

$$I_T = 14,000 \times 1.39 = 19,460 \text{ amperes}$$

$$I_R = 19,460 \div 1.1 = 17,69 \text{ amperes}$$

were

I_T = Total rms current at X/R = 50

I_R = rms symmetrical current at X/R = 17

The results above indicate that a standard circuit breaker having a preferred interrupting rating of 20 kA or higher should be selected.

Now, contemplating the case where for example, a standard 20 kA, 3 cycle circuit breaker is available and it is desired to apply this breaker on a system that has an X/R value of 80. The maximum interrupting capability of this circuit breaker for this application can be determined by simply multiplying the rated symmetrical capability by the ratio of the standard circuit S factor to the S factor corresponding to the high X/R system. The elapsed time, or contact parting time, for this circuit breaker is 35 ms. and the two S factors, from figure 9.4, are 1.2 and 1.56 respectively. The S factor ratio is then equal to 0.77 and the product of this factor times the symmetrical current rating is 20 × .77 = 15.4 kA which represents the new rated symmetrical current for the application on a system with an X/R of 80. If we were to assume that this circuit breaker was being installed in close proximity to a source of generation and if were to consider the effects of the transient ac component then the multiplying factor for the high X/R condition, as given in figure 9.5 would be approximately 1.42 and the ratio between factors is 0.84. The interrupting capability now becomes 16.9 kA. Comparing the results we see that the difference is within the range of accuracy of the circuit components and that in any case the error is on the safe side.

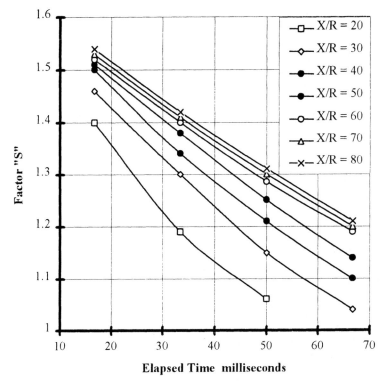

Figure 9.6 Asymmetrical factor "S" including ac decrement.

When evaluating the applications on systems that have a higher X/R than that assumed by the standards, the only consideration given so far is to the interrupting capability of the circuit breaker however, attention should also be given to the maximum current peak that can occur since this current peak is a function of the time constant, or X/R of the circuit and care must be taken not to exceed the maximum current peak that has been assigned to the circuit breaker. In figure 9.6 the multiplying factor for the peak currents is plotted as a function of the system's X/R and for the example given above we find that the current peak multipliers are 2.6 and 2.775 and the peak currents are 20 × 2.6 = 52 kA and in the worst case 16.9 kA × 2.775 = 46.9 kA for the standard rating and for the higher X/R, respectively.

9.3 Applications at Higher and Lower Frequencies

In general the question of how a circuit breaker will behave when applied at a higher or a lower frequency than that for which it was designed is posed mostly in applications involving medium voltage and up to 145 kV equipment. The most common low frequencies considered are those which are associated with transit applications the two most popular ones are 25 and 16 and 2/3 Hz. High frequency applications are rare and usually they are associated with very specialized applications in the medium voltage class.

Vacuum interrupters have demonstrated that it does not make any difference whether they are used on 50 or 60 Hz.. At lower frequencies, primarily, because of a lack of applicable data, it is customary to reduce the interrupting capability of the interrupter as a function of $I^2 \times t$.

The derating function is expressed as:

$$I_2 = I_1 \sqrt{\frac{f_2}{f_1}}$$

where

I_1 = Original symmetrical interrupting rating of circuit breaker
I_2 = Derated capability
f_1 = normal rated frequency (50 Hz)
f_2 = desired low frequency

For example for an application at 25 Hz. the short circuit current should be limited to 0.707 of the rated current at 50 or 60 Hz.

For applications of vacuum circuit breakers at frequencies higher than 60 Hz, experience has shown that the interrupter is not sensitive to the rate of change of current (di/dt) however, this does not mean that because of the shorter periods higher currents are allowable. The maximum short circuit current is still limited by the peak value of the current which is that of the original rating at 50 or 60 Hz.

For SF_6 circuit breakers the situation is different, as it was described in chapter 5, the interrupters, in most cases, are sensitive to the rate of change of current. This would imply that at lower frequencies higher currents can be interrupted; however, in the case of a puffer circuit breaker, even when the interrupter can successfully handle the extra input energy caused by the longer periods, it is unlikely that the speed and travel requirements can be properly accommodated by conventional designs. Therefore it is generally advisable that unless it is specifically sanctioned by the manufacturer, applications at low frequencies should be avoided. The same could be said for high frequency applications first; because of the derating required which is a function of the di/dt at the high frequency current and secondly, because the TRV would also increase as a function of the system frequency and by now we are cognizant of the high sensitivity of SF_6 interrupters to TRV values.

9.4 Capacitance Switching Applications

Capacitance switching applications involve not only interrupting capacitive currents, a subject which has been dealt in previous chapters, but also the energizing of overhead lines, cables and capacitor banks. Capacitor banks and cable systems may be either isolated, or back to back connected.

An isolated capacitor bank, or cable, is defined as follows [3]: "Cables and shunt capacitors shall be considered isolated if the maximum rate of change, with respect to time, of the transient inrush current does not exceed the maximum rate of change of the symmetrical interrupting current capability of the circuit breaker at the applied voltage."

This is represented mathematically by the following expression.

$$\left(\frac{di}{dt} \right)_{max} = \sqrt{2} \, \omega \, I \left[\frac{V_{max.(rated)}}{V_{applied}} \right]$$

where

$\left(\dfrac{di}{dt} \right)_{max.}$ = rate of change of inrush current

$\omega = 2\pi f$ or 377 for 60 Hz

f = power frequency

$V_{max.\ (rated)}$ = rated maximum voltage

$V_{applied}$ = maximum applied voltage

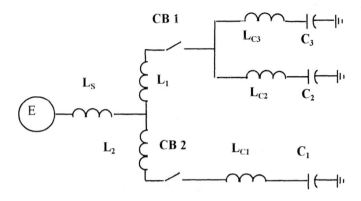

Figure 9.7 Single line diagram of a typical installation showing capacitor banks connected on a back-to-back fashion.

I = rated short circuit current

The following definition is given [3] for back to back capacitor banks or cable circuits: "Cable circuits and shunt capacitor banks shall be considered switched back to back if the highest rate of change of inrush current on closing exceeds that specified as the maximum for which the cable or shunt capacitor bank can be considered isolated." In simpler terms, isolated cable circuit, or capacitor bank means that only one cable, or one bank is on one bus, while back to back means that more than one cable, or capacitor bank is connected to the same bus. A typical system illustrating a back to back connection of capacitor banks is shown in figure 9.7.

9.4.1 Isolated Cable

Energizing a cable by closing the contacts of a circuit breaker will result in the flow of a transient inrush current. The magnitude and the rate of change of this inrush current is, among other factors, principally a function of the applied voltage, the cable geometry, the cable surge impedance and the length of cable.

Since it is given that the circuit breaker must be able to withstand the momentary short circuit requirements of the system the transient inrush current to an isolated cable is never a limiting factor in the application of a circuit breaker.

9.4.2 Back to Back Cables

When switching back to back cables high magnitude transient inrush currents that are accompanied by a high initial rate of change may flow between the cables being switched. The transient inrush current is limited by the surge imped

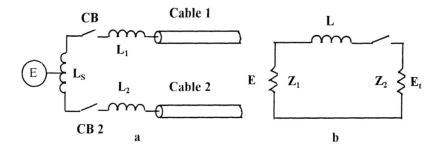

Figure 9.8 Back to back cable connected cables: **(a)** circuit single line diagram, **(b)** equivalent circuit for calculation of current.

where

E_m = peak of applied voltage
E_t = trapped voltage on cable being switched
Z_1, Z_2 = cable surge impedance
L = total circuit inductance between cable terminals
I_{MR} = rated peak inrush current

9.4.3 Isolated Shunt Capacitor Bank

The magnitude and the frequency of the inrush current resulting from energizing an isolated capacitor bank is a function of the point on the wave of the applied voltage where the contacts were closed, of the capacitance and inductance of the circuit, of the charge on the capacitor at closing time and of any damping resistance contained in the circuit.

The transient inrush current that flows into an isolated capacitor bank is less than the available short circuit current at the terminals of the circuit breaker and since the momentary current rating of the circuit breaker reflects the maximum short circuit current of the system then, the isolated capacitor bank inrush is of no consequence for the application of the circuit breaker.

When switching an isolated capacitor bank the value of the inrush current and its frequency is given by the following expressions.

$$i = \frac{E}{\sqrt{\dfrac{L}{C}}} \sin\left(\frac{t}{\sqrt{L_s C}}\right)$$

$$I_{peak} = \frac{\sqrt{2}E_{LL}}{\sqrt{3}} \sqrt{\frac{C}{L_s}} \quad \text{and}$$

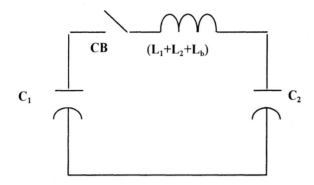

Figure 9.9 Simplified equivalent circuit for back to back capacitor banks calculations and where L_1 and L_2 are the inductance between capacitor banks and L_b is the bus inductance.

$$f = \frac{1}{2\pi\sqrt{L_s C}}$$

where

E_{LL} = line to line system voltage
L_s = system line inductance
C = bank's capacitance

9.4.4 Back to Back Capacitor Banks

When energizing capacitor banks in a back to back configuration high magnitudes and high frequencies of the inrush currents can be expected. The magnitude of the current being limited only by the impedance of the capacitor bank and by the inductance between the banks that are being energized.

A typical single line circuit representing the back to back condition is shown in figure 9.9 and the equations that define this condition are given below.

$$i = \frac{E}{\sqrt{\dfrac{(L_1 + L_2 + L_b)}{C_T}}} \sin\left(\frac{t}{\sqrt{(L_1 + L_2 + L_b)C_T}}\right)$$

$$I_{peak} = \frac{\sqrt{2}E_{LL}}{\sqrt{3}} \sqrt{\frac{C_T}{\left(L_1 + L_2 + L_b\right)}}$$

$$f = \frac{1}{2\pi\sqrt{\left(L_1 + L_2 + L_b\right)C_T}}$$

where

$C_T = C_1 + C_2$

L_1 and L_2 = Inductance between capacitor banks including inductance
of   the banks

L_b = inductance of the buss between the banks

The typical values for the inductance between capacitor banks are given in reference [4] and are reproduced in Table 9.1 below.

Table 9.1

Rated Maximum Voltage (kV)	Inductance per Phase of bus (μH/ft)	Typical Inductance between banks (μH)
15.5 & above	0.214	10 - 20
38	0.238	15 - 30
48.3	0.256	20 - 40
72.5	0.256	25 - 50
121	0.261	35 - 70
145	0.261	40 - 80
169	0.268	60 - 120
242	0.285	85 - 170

9.4.5 General Application Guidelines

The first factor that needs to be taken into account when considering the application of a circuit breaker for capacitance switching is the type of circuit breaker to be used. If, by any chance, the candidate is an oil circuit breaker, then very serious consideration must be given to avoid the possibility of exceeding the given rating for the maximum frequency of the inrush current. Oil circuit breakers are extremely sensitive to the inrush current and failing to operate below the maximum limits could lead to catastrophic failures.

With modern circuit breakers the frequency of the inrush current is a lesser concern for the circuit breaker itself; however, in most cases still constitutes the limiting factor because of other equipment in the system such as linear couplers and current transformers and also due to the effects of the induced voltages on the control wiring and the possible rise on ground mat potentials.

The problem with linear couplers and current transformers is related to the secondary voltage that is induced across the terminals and which can be calculated using the following formulas:

For linear couplers

$$E_s = \frac{f_2}{f_1} \times \text{LCR} \times I$$

For current transformers

$$E_s = \frac{f_2}{f_1} \times \text{CTR} \times I \times L_{relay}$$

where

E_s = secondary voltage across device terminals
LCR & CTR = linear coupler or current transformer ratio
f_1 = system power frequency
f_2 = transient frequency
I = transient current
L_{relay} = relay's inductance ≈ 0.3 ohms

Just for illustration purposes lets us assume an inrush current of 25 kA and an inrush frequency of 6400 Hz; for a linear coupler and a current transformer both having a ratio of 1000 to 5, then the calculated secondary voltage are:

$$\text{Linear coupler } E_s = \frac{6400}{60} \times \frac{5}{1000} \times 25000 = 13,330 \text{ volts}$$

$$\text{Current Transformer } E_s = \frac{6400}{60} \times \frac{5}{1000} \times 25000 \times 0.3 = 4000 \text{ volts}$$

The results from this example indicate that voltage limiters need to be used to protect either the transformer, or the linear coupler.

The magnitude of the peak current, as long as it does not exceed the maximum peak of the given close and latch capability of the circuit breaker, should not present any problem even if its greater than the values that are published as standard ratings. Nevertheless before exceeding the rating values it should be verified with the manufacturer of the circuit breaker that there are no design features that may say otherwise.

9.4.5.1 Limiting Inrush Frequency and Current

When found that it is necessary to limit the magnitude and the frequency of the inrush current what is recommended is the use of:

1. Closing resistors or inductors, which are inserted momentarily during the capacitor energizing period and then are subsequently bypassed.

2. Fixed reactors which are permanently connected to the circuit. This procedure reduces the efficiency of the capacitors and increases the losses of the system

3. Synchronous switching where the closing of the contacts is synchronized so that it takes place at or very near the zero voltage thus effectively reducing the inrush current. The poles of the circuit breaker for this operation must be staggered by 2.7 milliseconds.

9.4.5.2 Application of Circuit Breakers Near Shunt Capacitor Banks

Special care must be taken to insure that line circuit breakers are not applied to switch capacitor banks because for certain types of faults the contribution made to the fault by the out-rush current from the capacitor banks will expose the circuit breaker to currents that are, in most cases, greater than those encountered on back to back switching, which means that not even those breakers rated as definite purpose circuit breaker may be suitable for these applications. The solution to this problem may the inclusion of pre-insertion and opening resistors on the circuit breakers or the installation of current limiting reactors in series either with the capacitor banks, or with the individual circuit breakers. A single line diagram illustrating a circuit configuration leading to large amounts of out-rush currents is shown in figure 9.10.

9.5 Reactor Current Switching, High TRV Applications

In general switching of reactor currents is associated with small magnitude of currents, high frequency transient recovery voltages, and high overvoltages. It is then conceivable that in some of those applications involving reactor switching, when current limiting reactors are connected in a close proximity to

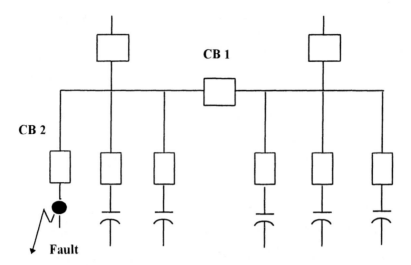

Figure 9.10 Diagram illustration of typical installation where the fault contribution of the capacitor banks may produce currents that exceed the capabilities of circuit breaker CB2.

the circuit breaker, the resulting TRV may exceed the limits for which the circuit breakers have been designed and tested.

As it should be recalled SF$_6$ circuit breakers are likely to be more limited in their capability to withstand higher rates of recovery voltage than similarly rated vacuum circuit breakers. Consequently for these applications, and if available, a vacuum circuit breaker could be a better choice. Nevertheless, and since we also know that at voltages higher than 38 kV the most likely choice would be an SF$_6$ circuit breaker. One solution to reduce the rate of rise of the recovery voltage so that the circuit breaker is not overstressed, specially during the thermal recovery period which takes place during the first 2 microseconds after current interruption is to add capacitors, either in parallel to the interrupter contacts, or connected from line to ground at the terminals of the circuit breaker. For this simplistic approach the size of the capacitor can be calculated by simply assuming that the TRV is produced by an equivalent series LC circuit and where the initial time delay (t_d), that is now required has to be greater than 2 microseconds, is given approximately by:

$$t_d = Z \times C_m \text{ in microseconds}$$

where

$$Z = \text{surge impedance}$$

C_m = externally added capacitance in microfarads

It is also possible, and this is a more realistic approach, specially when dealing with applications at the higher end of the voltage scale, to utilize circuit breakers that are equipped with opening resistors, or to use metal oxide surge arresters applied directly to the circuit breaker.

Opening resistors are connected in parallel to the main interrupting contacts of the circuit breaker and constitute an effective method for the reduction of overvoltages and for the modification of TRV. The value of the closing resistor should be approximately equal to the ohmic reactance of the reactor.

Another approach that can be used and which will be discussed in more detail in the next chapter is the synchronized opening of the circuit.

When the circuit breaker is used to switch shunt reactors that are connected to the bus its fault current interrupting capability should be determined in relation to the full requirement of the system but, if the circuit breaker is used to switch shunt reactors that are connected to transmission lines the full fault capability may not be required although the short time and the momentary capabilities of the circuit breaker should be at least equal to the ratings of the circuit breaker that is providing the primary fault protection for the circuit.

9.5.1 Ferroresonance

As described above the use of capacitors helps to improve the TRV withstand and therefore the interrupting capability of a circuit breaker, in other instances capacitors are also placed across the contacts on poles that have multiple interrupters with the purpose of equalizing the voltage distribution across the individual interrupters. In general, for these purposes the larger the capacitance the greater the improvement; however, the down side to the use of larger capacitors, aside from the cost and added complexity, is the possibility of creating a ferroresonant condition between the capacitor and the potential transformers that may be connected to lines that are de-energized.

This condition is created when there is a transformer connected to the bus, in which case there is a series connection of the capacitors and the transformer as illustrated in figure 9.11. The equivalent circuit, basically represent a voltage divider $(X_m / X_c - X_m)$ and when the capacitive reactance X_c equals the inductive reactance X_m of the transformer then, at least in theory the voltage at the bus will become infinite, but in reality the voltage is limited due to the non linear impedance of the transformer and its magnitude is determined by the intersection of the capacitive voltage with the transformer's saturation voltage and whenever the intersection point is below the knee of the saturation curve then the overvoltage could be severe and damage to the transformer can take place. An expeditious solution for this problem is to add a low ohmic resistor connected across the secondary of the transformer.

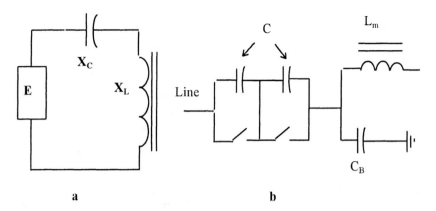

Figure 9.11 Relationship of circuit components for ferroresonance: (a) equivalent circuit and (b) single line diagram.

9.6 High Altitude Dielectric Considerations

The application of a circuit breaker at elevations greater than 1000 meters (3300 ft) dictates that its dielectric capabilities be reduced due to the lighter air density. For sealed interrupters, the withstand capability across the contact gap is not affected and only those insulating structures which are exposed to the air atmosphere should be considered for derating. Keeping in mind that what is desirable, but not always attainable, is that in case of a flashover due to excessive overvoltages, said flashover should occur externally to the contacts. Depending on the type of interrupter used and on the design of the circuit breaker it is possible that there is air dependent insulation located in a parallel path to the SF_6 or vacuum insulation across the breaker contacts. This implies that in these cases the altitude derating must be applied.

In those cases where the design provides sufficient coordination between the non-atmospheric and the atmospheric paths it is conceivable that the possibility of applying the equipment either without derating or with a limited derating may be considered; however this approach must be carefully evaluated to assure that the insulation coordination with the rest of the equipment involved on the particular installation is not compromised in any way. Furthermore it will be necessary to provide adequate protection for the circuit breaker in the form of properly rated arresters located both at the line side and at the bus side and this may prove to be uneconomical when compared to a fully rated circuit breaker.

The applicable derating factor (K) is given by the following equation and is shown in figure 9.12.

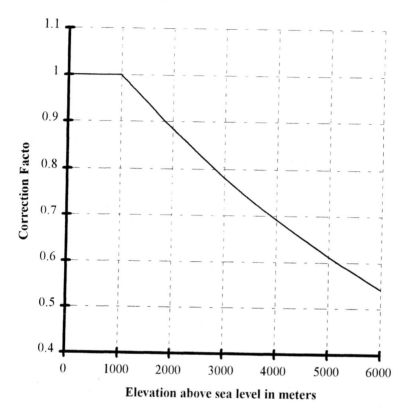

Figure 9.12 Correction factor for the dielectric withstand of the components of a circuit breaker exposed to air ambient at elevations other than sea level.

$$K = e^{-\left((H-1000)\big/_{8150}\right)}$$

where

K = derating factor

H = altitude where circuit breaker is to be applied in meters

Once a correction factor has been determined, the next step is to calculate the operating voltage rating at the standard conditions. The selection of a circuit breaker that has a rating equal or greater than that which has been calculated will generally take care of the power frequency and the impulse dielectric withstand requirements at the new altitude. For example, let us chose a circuit breaker for a 145 kV system to be used at 3000 meters above sea level.

The factor K is equal to $e^{-(3000-1000/8150)} = 0.782$

The maximum operating voltage rating at standard conditions for this application is equal to $1/\ 0.782 \times 145 = 185$ kV. The closest higher standard rating is 242 kV.

Now let us suppose that the maximum voltage of the system is 121 kV instead of 145 kV what we find with these new conditions is that the required maximum operating voltage is 155 kV. Now we are faced with a situation where it may be possible to use a 145 kV circuit breaker protected by properly sized arresters or to opt for selecting a 242 kV circuit breaker. If a dead tank circuit breaker is being considered and the system is grounded then the recommended choice should be the 145 kV rated circuit breaker.

9.7 Reclosing Duty Derating Factors

The need for derating factors for applications involving rapid reclosing or extended duty cycles depends primarily on the type of circuit breaker that is being used. Being more specific, it should apply only to oil or air magnetic circuit breakers. This is due to the fact that following an interruption these circuit breakers need additional time to cool off and to regain their dielectric capability. SF_6 and vacuum circuit breaker generally do not need the extra time and therefore there should be no need for reducing their interrupting capability; nevertheless, there may be some cases where the manufacturer chooses to do so and consequently when the application of these breakers involves either more operations, or shorter time intervals between operations than what is established as the standard duty cycle it is advisable to consult with the manufacturer.

When a derating factor needs to be applied it can be calculated used the relationship that is given below, and which is applicable within the following constrains.

1. the duty cycle does not consists of more than operations
2. all operations taking place within a 15 minute period are to be part of the same duty cycle.
3. a period of 15 minutes between operations is considered to be sufficient for the initiation of new duty cycle.

The derating factor is obtained by first calculating the total required per cent reduction factor D which is found by adding the individual reductions that are obtained by multiplying the constant d_1 given in figure 9.13 by the number

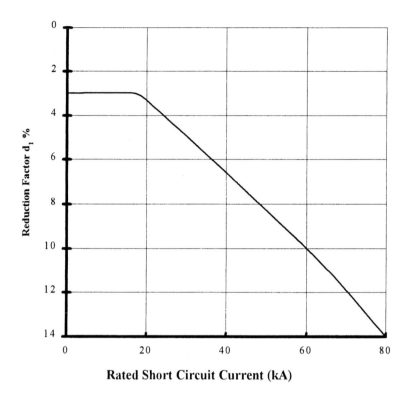

Figure 9.13 Reduction factor d$_1$ that is used for calculating derating factors for rapid reclosing and extended duty cycles.

of operations in excess of the two which are required by the standard duty cycle (CO + 15 sec + CO) and by the ratio of the time difference for each reclosure with a period that is less than 15 seconds. This is expressed mathematically by the relationship given below:

$$R = 100 - D$$

$$D = d_1(n-2) + d_1 \frac{(15 - t_1)}{15} + d_1 \frac{(15 - t_2)}{15} + \text{\_\_\_\_\_}$$

where

        R = derating factor

        D = total reduction factor in percent

        d_1 = multiplier for the calculation of reducing factor

t_1 = first time interval which is less than 15 seconds

t_2 = second time interval that is less than 15 seconds

n = total number of contact openings

The following example should serve to illustrate the concept. Given a 245 kV, 20 kA circuit breaker that is going to be applied for a (O + CO + 10 sec + CO + 1 minute + CO) duty cycle.

The multiplier d_1 from figure 9.13 is equal to 3.3 at 20 kA and the reduction factor D is

$$D = 3.3 \times (4 - 2) + 3.3 \times \frac{(15 - 0)}{15} + 3.3 \times \frac{(15 - 10)}{15} + 0$$

$$D = 6.6 + 3.3 + 1.1$$

$$D = 11\% \text{ and } R = 100 - 11 = 89\% \text{ or } 0.89$$

The short circuit rating of the circuit breaker is then reduced to 20 x .89 = 17,800 amperes

9.8 Choosing Between Vacuum or SF$_6$

A fair comparative assessment of vacuum or SF6 circuit breakers can only be made for medium voltage circuit breakers where both types of technologies can be used interchangeably. The choice between vacuum, or SF$_6$ is mostly a matter of preference. The basic performance of both technologies is essentially the same because both are designed and tested to meet the same performance standards. There may be however some specific applications where one technology may be deemed more appropriate.

Vacuum circuit breakers have a very long and relatively maintenance free life which represents a desirable attribute and a significant advantage for this technology. The main disadvantage for vacuum interrupters, on the other hand, is their noticeable propensity for initiating overvoltages which may be harmful to other equipment. Although for most applications there is no need for concern, it is recommended that for applications involving the switching of transformers and/or rotating machinery due consideration be given to the use of surge arresters, and better yet to the use of surge suppressers which consist of a resistor and a capacitor in series. This components combination not only reduces the frequency of the transient voltage but it also reduces the magnitude of the voltage. Another advantage of this protection is that it serves to detune the circuit and prevents the possibility of having a resonant circuit.

For applications where a large numbers of operations under load, or fault conditions are required or where high rates of rise of recovery voltage are expected such as in the case of reactor switching, vacuum circuit breakers may be the better choice. But in the other hand for applications on capacitor switching or the switching of transformers SF_6 circuit breakers may be advantageous. In either of the last two mentioned applications the addition of capacitors or surge suppresser will serve to equalize the applicability of both technologies. Another factor that may influence the choice and which at the time of this writing is still unknown are the possible future environmental restrictions and liabilities that may be imposed on the use of SF_6.

REFERENCES

1. ANSI/IEEE C37.24-1986 IEEE Guide for Evaluating the Effect of Solar Radiation on Outdoor Metal-Enclosed Switchgear.
2. ANSI/IEEE C37.010-1979 IEEE Application Guide for AC High-Voltage Circuit Breakers Rated on a Symmetrical Current Basis.
3. ANSI/IEEE C37.04 Rating Structure for AC High-Voltage circuit Breakers Rated on a Symmetrical Current Basis.
4. ANSI/IEEE C37.012 Application Guide for Capacitance Current Switching for AC High-Voltage Circuit Breakers Rated on a Symmetrical Current Basis.
5. ANSI IEEE C37.015 IEEE Application Guide for Shunt Reactor Switching.
6. D. L. Swindler, Application of SF_6 and Medium Voltage Circuit Breakers, IEEE CH 3331-6/93/0000-0120, 1993.
7. V. D. Marco, S. Manganaro, G. Santostino, F. Cornago, Performance of Medium Voltage Circuit Breakers Using Different Quenching Media, IEEE Transactions on Power Delivery, VOL-PWRD-2, No. 1, Jan 1987.
8. J. H. Brunke, Application of Power Circuit Breakers for Capacitive and Small Inductive Current Switching, IEEE Tutorial Course, Application of Power Circuit Breakers, 93-EHO-388-9-PWR, 43-57, 1993.
9. S. H. Telander, M. R. Willheim, R. B. Stump, Surge limiters for Vacuum Circuit Breaker Switchgear, IEEE Transactions on Power Delivery Vol. PWRD-2 No. 1 Jan. 1987.
10. A. N. Greenwood, D. R. Kurtz, J. C. Sofianek, A guide to the application of Vacuum Circuit Breakers, IEEE Transactions PA & S Vol. PAS-90, 1971.

10

SYNCHRONOUS SWITCHING AND CONDITION MONITORING

10.0 Introduction

Synchronous switching and condition monitoring are two subjects that have gained a great deal of relevance not only because of their potential for increasing reliability and for making a contribution to improve the overall power quality of the electric systems, but also for economic reasons. These concepts can be instrumental in minimizing the use of auxiliary components, such as pre-insertion resistors, in reducing equipment wear and unnecessary maintenance and thus reducing the total cost of ownership throughout the full life time of the equipment.

Synchronous switching is not a recent, or a new idea and for at least the last 30 years the feasibility of synchronized switching has been studied; many concepts have been investigated and even some commercial equipment has been built and utilized. [1, 2, 3]

Condition monitoring is a relatively newer concept that has came about primarily because of recent developments of electronic sensors and data acquisition equipment that have made this idea not only technically feasible but also economically attractive. Condition monitoring is an essential component for synchronous switching simply because a lot depends on how accurately the operating characteristics of the circuit breaker can be controlled. It is well know that the operating characteristics can be affected by extreme ambient temperatures, and by other prevailing conditions such as mechanism operating energy levels, control voltages, operating frequency of the equipment, its chronological age, and its maintenance history among others. Collecting information about these variations would provide with a data source from where suitable correction factors may be selected to compensate for those operating deviations which are critical for accurate synchronous operation.

10.1 Synchronous Switching

Opening or closing the contacts of a circuit breaker is normally done in a totally random fashion and consequently, as it has been described before, transient current and voltage disturbances may appear in the electrical system. A typical way for controlling this transient behavior has been to add discrete components such as resistors, capacitors, reactors, surge arresters, or combinations of the above to the terminals of the circuit breaker nevertheless, in many cases it would be possible to control these transients without the addition of external components but by operating the circuit breaker in synchronism with either the current or the voltage oscillations, depending upon the switching operation at hand. This means that for example, the opening of the contacts should occur at a current zero when interrupting short circuit currents, or that the closing of the contacts should take place at voltage zero when energizing capacitor banks.

Operations that can benefit the most by synchronized switch are those involving the switching of unloaded transformers, capacitor banks, and reactors. Energizing transmission lines and opening the circuit breaker to interrupt short circuit currents are also good candidates for synchronous switching [4].

Ideally and to obtain the greatest benefits, synchronous switching should be done using circuit breakers that are capable of independent pole operation. Independent pole operation is already a standard feature in circuit breakers that are rated above 550 kV and it is also used, under special request, for applications as low as 145 kV. However for those designs where all three poles are operated in unison the implementation of controlled switching concepts will require the development of specially designed circuit breakers which are provided with suitable methods for staggering the pole operating sequences.

To select the operating characteristics of a circuit breaker which have the most direct impact for synchronized switching requires a clear definition and understanding of the cause and effect relationship that exists between the mechanical operation of the circuit breaker and the behavior of the electric system. In every case it is indispensable to closely analyze the mechanical and electrical properties of the design, including contact velocity, contact opening and closing time, minimum arcing time for different interrupting duties and current levels and cold gap voltage withstand capability; furthermore and in relation to the operating times the effects of control voltage fluctuations, ambient temperature, tolerances of the mechanism's stored energy and operating wear must also be considered.

10.1.1 Mechanical Performance Considerations

Consistency in the making and breaking times of the circuit breaker is absolutely essential for successful implementation of all types of synchronous switching. But considering the fact that a circuit breaker is a mechanical device, an even though modern designs are highly reliable, they must still be improved to minimize the influence of wear, aging, cold temperatures, and voltage control sources. Of all these parameters the ones that have the greatest influence are the ambient temperature, the level of energy stored in the operating mechanism and the control voltage level.

The effects of wear and aging upon the equipment presently in service, at least for now, are unpredictable. There is only scarce and incomplete data and furthermore there is a need for data obtained from actual field operating experiences to fully evaluate the influence of these conditions upon the life of a circuit breaker. Another condition that is difficult to evaluate and which may have a significant influence upon the repeatability of the operations is the effects of long periods of time when the breaker has not been operated, specially, when during the idle periods the equipment has been subjected to very low ambient temperatures.

TABLE 10.1

Typical Operating Characteristics
of an EHV Puffer Circuit Breaker

	CLOSING		OPENING	
	Time. m/s	Speed m/s	Time m/s.	Speed m/s
Normal. Control Volts	73.9	4.3	16.0	8.6
Min. Control Volts	105.4	4.0	17.1	8.4
Max. Control Volts	72.1	4.0	15.9	8.4
Min. Temp. -40°C	76.4	5.0	17.2	8.5
Min. Force Mech.	76.4	3.9	16.6	7.6
After 5000 Operations	73.9	4.3	16.0	8.6

TABLE 10.2

Typical Operating Characteristics
of a Vacuum Circuit Breaker

	CLOSING		OPENING	
	Time. m/s	Speed m/s	Time m/s.	Speed m/s
Normal. Control Volts	45	0.9	25	1.25
Min. Control Volts	49	0.9	27	1.25
Max. Control Volts	43.5	0.9	23.5	1.25
Min. Temp. -40°C	46.5	0.85	26	1.2
Min. Force Mech.	48	.8	26	1.2
After 5000 Operations	47	.9	25	1.25

As an illustration only and for future reference, the operating characteristics of what may be considered, from a qualitative point of view, typical of a SF$_6$ puffer high power circuit breaker are tabulated below in table 10.1. Similar values for vacuum interrupters are given in table 10.2.

·In addition to the mechanical characteristics, shown in tables 10.1 and 10.2 the following electric attributes representing the minimum arcing times for different types of switching duties are given.

Puffer SF$_6$

Short Circuit Current Interruption	13 milliseconds
Capacitive Bank Current Interruption	2.5 milliseconds
Low Inductive Current (Reactors) Interruption	3 milliseconds

Vacuum

Short Circuit Current Interruption	4 milliseconds
Capacitive Bank Current Interruption	1.5 milliseconds
Low Inductive Current (Reactors) Interruption	2 milliseconds

For most synchronous switching applications the aggregate of all the variations in the operating times, corresponding to the worst condition, both in closing and in opening, should not exceed a maximum of plus or minus one millisecond. For example, observing the operating times that are shown in table 10.1, it is evident that the aggregate time of this circuit breaker, as shown below, far exceeds the maximum opening or closing time allowable deviation.

Maximum deviation in closing time

31.5 (min. volts) + 2.5 (min. temp) + 2.5 (min. force) = 36.5 ms

Maximum deviation in opening time

1.1 (min. volts) + 1.2 (min. temp) + 0.6 (min. force) = 2.9 ms

It is also quite evident that the greatest influence on the operating times is exerted by the control voltage. This parameter, in comparison to all the others offers better possibilities for enhancement and consequently it is where the major improvement would be expected. Possible solutions are the use a regulated power supply, or capacitor discharge systems for the supply of the control circuits energizing the operating coils. In addition the solenoid coils should be optimized to reduce the operating time range.

So far mention has only been made of the operating times without mention of the operating speeds, but, these also are quite important for they are related to the minimum arcing time during opening and to the prestrike time during closing. The product of the velocity and the above mentioned parameters determine the minimum contact gap required for the respective operation.

10.1.2 Contact Gap Voltage Withstand

The contact gap voltage withstand capability is an arbitrary definition that relates to the change in gap distance in relation to the instantaneous voltage withstand capability across said gap during a normal closing operation. An approximate value can be obtained for an interrupter, but first, to minimize the variables, it is assumed that the interrupter has not been subjected to an electric arc immediately preceding the measurement. The voltage withstand thus obtained is somewhat in the optimistic side, since during normal service, and more so during a reclosing operation it will be reasonable to expect that arcing has occurred. The effects of prior arcing must still be investigated, but it is expected that if there is a reduction in capabilities such reduction would be less than 10% of the cold gap withstand.

For SF_6 interrupters the contact gap withstand is a function of the gas pressure and of the geometry of the contacts; higher operating pressures and lower dielectric stresses in the field across the contacts would produce higher contact gap withstand values. Reported values [5] range from approximately 10 kV per millimeter to 25 kV per millimeter. While for vacuum interrupters the contact gap capability [6] is primarily a function of the electrode material and is in the range of 20 to 30 kV per millimeter.

10.1.3 Synchronous Capacitance Switching

For capacitance switching, the primary concern is not as much the interruption of capacitive currents because, due to the inherent characteristics of vacuum and SF_6 circuit breakers the problems associated with restrikes, found with earlier technologies, have been greatly reduced and today, indeed restrikes are a very rare occurrence. On the other hand, failures are often reported which are the direct result of inrush currents and overvoltages that have propagated themselves into lower voltage networks causing damage specially to electronic equipment connected to the circuit. A comparison of the voltage transient for a non-synchronous operation is shown in figure 10.1 and in figure 10.2 the voltage response of a synchronized closing is shown. As it can be seen in the illustrations the higher frequency component of the voltage is practically eliminated when the contacts are closed at a nominal voltage zero condition.

Energizing a capacitor bank. In order to completely eliminate the overvoltages produced by the closure of a circuit breaker onto a capacitor bank it is required that there be a zero voltage difference across the contacts of the circuit breaker; naturally this is not always possible simply because some deviation from the optimum operating conditions is to be expected. Studies [7] have shown that the overvoltages can be reduced to acceptable limits when the closing of the contacts occur within one millisecond either before or after the voltage zero point. The significance of this requirement is better appreciated when consid-

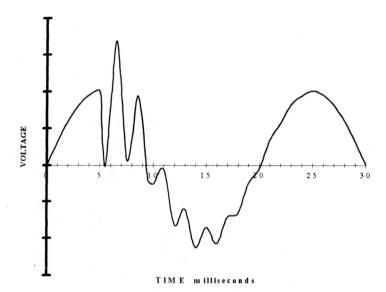

Figure 10.1 Voltage corresponding to a non-synchronous closing of a circuit breaker into a capacitor bank.

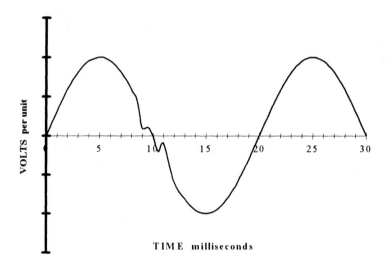

Figure 10.2 Voltage corresponding to a circuit breaker synchronous closing into a capacitor bank

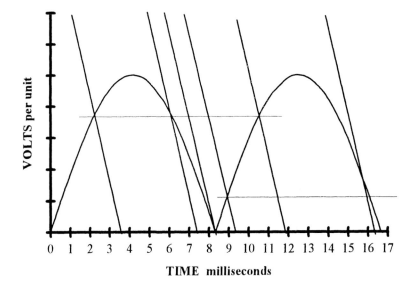

Figure 10.3 Relation between system voltage and interrupter gap withstand capability.

ered in conjunction with the gap withstand capability of the contacts as illustrated in figure 10.3 where the absolute value of a sinusoidal voltage is plotted in conjunction with the slope of an assumed gap withstand characteristic. As it can be seen in the figure the point where the prestrike takes place corresponds to the intersection of the two curves. In the figure to better illustrate the concept the several different times for the zero intersection of the gap withstand are shown. Ideally the rate of change of the gap withstand voltage should be at least 10% higher than the rate of change of the system voltage to assure proper coordination of the two rates..

To further investigate the concept let us consider the circuit breaker whose characteristics were shown in table 10.1, let us furthermore assume that the gap withstand capability of this circuit breaker is 10 kV per millimeter. Using the given closing speed of 4.3 meters per second, shown in table 10.1, it is found that the corresponding rate of change of withstand capability is 43 kV per microsecond.

Assuming that the circuit breaker under consideration is intended for capacitance switching duty on a 72 kV ungrounded system which. Since, as we know, an ungrounded system represents the worst case in terms of voltage peak because this peak, for the last phase to close, is equal to 1.5 times the peak of the rated line to line system voltage.

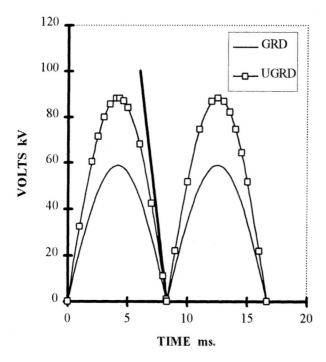

Figure 10.4 Rates of change of gap withstand capability for the circuit breaker of the given example and the rates of change of the system voltage for grounded and ungrounded applications.

The maximum rate of change of voltage with respect to time at the instant of voltage zero corresponding to the above phase is

For a grounded application

$$\frac{dE}{dt} = \frac{\sqrt{2}}{\sqrt{3}} E \times \omega = 0.82 \times 72 \times 377 = 22.2 \text{ kV per microsecond}$$

For an ungrounded application

$$\frac{dE}{dt} = 1.5 \times \frac{\sqrt{2}}{\sqrt{3}} E \omega = 1.225 \times 72 \times 377 = 33.2 \text{ kV per microsecond}$$

Comparing the two rates of change, the ones corresponding to the system voltage against the one related to the rate of change of the gap capability, as it shown in figure 10.4, it appears that the interrupter being considered in the example would be adequate for this application.

Table 10.3

Rated Voltage	Maximum rate of change of system Voltage kV/μs	
	Grounded Bank	Ungrounded Bank
72	22	33
121	37	56
145	45	68
245	76	114
362	112	168
550	169	254

However, in the event that the example interrupter is considered for an application at 145 kV; by following the same procedure as before and using the tabulated values shown in table 10.3 for the customary preferred rated system voltages it is evident that this interrupter is inadequate for the application. Nonetheless, if the gap withstand capability is increased to 25 kV per millimeter, assuming the same 4.3 meters per second speed, then the rate of change of the withstand capability becomes approximately 108 kV per microsecond. This suggests that it would be possible to consider the use of this interrupter for all grounded capacitor bank applications up to 245 kV and if two of these interrupters are connected in series it would also be possible to meet the 550 kV application requirements. For ungrounded banks a single interrupter is good only up to 145 kV, and even with two interrupters in series it would not be possible to meet the 550 kV rating.

To increase the rate of change of the withstand capability in an SF_6 circuit breaker any of the following three options either individually or in combination may be applied.

1. increase the gas operating pressure
2. reduce the electric field stress in the contact region
3. increase the closing velocity of the circuit breaker contacts

Increasing the gas operating pressure, in many cases, is not a viable solution because of the possibility of SF_6 liquefaction at low temperatures. Nevertheless, it should be kept in mind that at voltages above 362 kV the systems are grounded almost without exception and therefore our imaginary non-ideal circuit breaker may already be acceptable.

For vacuum circuit breakers, and since presently they are used almost exclusively at system voltages in the range of 15 kV to 38 kV, the minimum rate of change of the gap withstand does not present any problem. The minimum rate of gap dielectric may be assumed to be in the neighborhood of 30 kV per millisecond, while the maximum rate of change of voltage for a 38 kV ungrounded bank is 17.5 kV per millisecond.

The precise points where the contacts of each pole must close depend upon the system connections. When the capacitors and the system neutrals are grounded then the optimum point to close the contacts is each pole independently at the voltage zero of the corresponding phase.

When the capacitors are connected in an ungrounded system it is possible to close the first pole at random, since there will be no current flow with only one pole closed. The second pole and the third pole must then close at their respective voltage zero. Another alternative would be to close two poles simultaneously at a voltage zero and then close the third pole at its corresponding voltage zero.

When we speak of voltage zero what is mean is that the voltage difference across the contacts of the circuit breaker is zero. Because of the possibility of trapped charges in a capacitor it would then be necessary to monitor the ac voltage from the source side as well as the dc voltage in the capacitor at the load side of the circuit breaker.

Presently, there are a number of control devices available. As an example of one such device, which is marketed under the name ACCUSWITCH, its block diagram is shown in figure 10.5.

De-energizing a capacitor bank. Capacitive currents generally require very short arcing times which means that the actual contact gap is very short and in some cases, when the magnitude of the recovery voltage exceeds the dielectric capability of this small gap, it leads to restrikes.

It was indicated earlier that typical minimum arcing times for capacitance switching with SF_6 circuit breakers is about 2.5 milliseconds and 1.5 millisecond for vacuum interrupters. Since it is not indispensable that the opening of the contacts coincides with the minimum arcing time but rather that it should be longer than that which is considered minimum, a satisfactory choice would be to part the contacts at a point that is at least 4 milliseconds prior to the current zero. Consequently the controls for this type of application need not be that sophisticated again all that is needed is that the contacts separate sufficiently ahead of the current zero.

As it was said before, with the advent of the new technologies of high voltage circuit breakers there is basically no need for synchronous opening of the capacitor banks and that this mode of operation should only be considered in very special occasions when it is known that there is a real possibility of restriking.

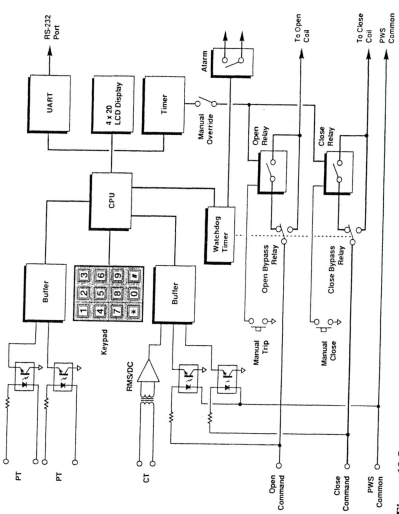

Figure 10.5 Schematic diagram of control device made by ACCUSWITCH.

10.1.4 Synchronous Reactor Switching

For reactor switching operations the basic needs are the opposite of those considered to be desirable for capacitance switching, that is closing the circuit is not as important as is opening.

Closing Control. Typically, most high voltage circuit breakers will pre-strike during a closing operation and as a result of this pre-strike an overvoltage that generally is less than 1.5 per unit is produced. Since this overvoltage, by no means, should be considered to be excessive there is no pressing need to control its magnitude and consequently a controlled, or synchronized closing, from the point of view of switching overvoltages is considered to be unnecessary. Furthermore if the closing is synchronized with a voltage zero condition this would result in a high asymmetric current which may develop excessive mechanical stresses within the turns of the reactor being switched on. Additionally, if in a grounded circuit the closing of the contacts takes place at a voltage zero it is possible that an excessive zero-sequence current may flow thus raising the possibility of the zero-sequence relays being activated.

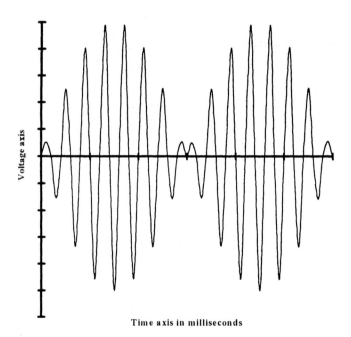

Figure 10.6 Representation of voltage appearing across the circuit breaker contacts when reclosing into a shunt compensated line.

If synchronous closing is to be considered, it would be preferable to close the contacts at maximum voltage, which makes the task relatively easy since there is a natural tendency for the contacts to do just that, plus the fact that near the peak of the voltage its rate of change is basically zero and therefore there is room for a larger tolerance in the allowable variation of the closing time.

One simple way to accomplish controlled closed would be to reduce the closing speed of the contacts so that the rate of change of the gap withstand becomes significantly lower than the maximum rate of change of the system voltage.

A unique condition that is worth mentioning because of the significant benefits that can be attained from synchronous closing, is when rapid reclosing of a shunt reactor compensated line is required. Reclosing implies that current interruption has just taken place and since following interruption a trapped charge may be left on the unloaded line. In this case the voltage across the circuit breaker will show a significant beat, as illustrated in figure 10.6, due to the frequency difference between the line and the load sides. At the source side of the circuit breaker the voltage oscillates with the power frequency while at the load side the frequency of the oscillation may be as low as one-half that of the power system frequency (60 Hz.) or as high as to approach the power frequency, it only depends upon the degree of compensation; the higher the compensation, the lower the frequency.

Since synchronization should be made at a beat minimum, where the voltage across the contacts is relatively small, it is evident that the degree of complexity for detecting this zero voltage condition across the contacts has greatly increased, thus making this task extremely difficult. This is further complicated by the fact that the variable beat frequency creates a high degree of uncertainty for predicting the voltage zero across the circuit breaker contacts.

Opening Control. Synchronized opening of the contacts in an application involving the switching of reactors should be considered for the purpose of reducing overvoltages that may be generated as the result of current chopping or reignitions that may occur during a normal opening operation. One benefit of synchronous opening of reactor circuits, specially those that use reactors for shunt compensation, is that it substantially reduces switching surge overvoltages.

Synchronous control for opening reactor circuits is not difficult to achieve since it is only necessary to separate the contacts at a time which is larger than the minimum arcing time required for that operation by that particular circuit breaker design. What is important is that the contact gap be sufficient to withstand the recovery voltage and that the contact separation be close enough to the minimum arcing time to reduce the possibility of current chopping.

The likelihood of reignitions is greatly reduced by synchronized switching even when the three poles of the circuit breaker are gang operated. If the poles are operated independently of each other and each allowing an arcing time of at least 4 milliseconds then any probability of reignitions is virtually eliminated.

It should be noticed that the synchronizing requirements for this type of applications are dependent primarily upon the characteristics of the circuit breaker, rather than the characteristics or the connections, grounded or ungrounded, of the system.

10.1.5 Synchronous Transformer Switching

Basically speaking, the switching of an unloaded transformer is no different than switching a reactor, that is the voltages and currents involved in opening and closing the circuit of the transformer generally have the same characteristics of those produced by the switching of reactors. However, for this application the most critical variable is the transformer's inrush current [8] which, in some occasions can reach magnitudes that approach those of the short circuit current. The magnitude of the inrush current depends on the transformer's impedance, on the magnetic characteristics of the core of the transformer and on the status of its magnetic flux remnants at the instant when the circuit is energized. The severity of the energizing process is greater for transformers that have a high remnants, than for those that are completely demagnetized. It follows then that for full synchronization it is necessary to detect the remnants level prior to the energizing of the transformer, or alternately that all openings of the transformer circuit be made synchronously so that the remnants conditions are controlled and can be well defined for the next closing operation. If remnants is not considered to be a problem then closing may be done satisfactorily at voltage peak with a tolerance of as much as ±2 milliseconds.

10.1.6 Synchronous Short Circuit Current Switching

Synchronized switching of short circuit currents is a desirable feature from the point of view of reducing contact erosion which translates in extending the life of the circuit breaker. However the benefits need to be weighted against the possible cost and the complexity of the task.

Synchronous closing. The aim of synchronous closing would be to reduce contact erosion by reducing the arcing time during closing, which is due to pre-strikes across the contacts. The benefits that may be achieved by synchronous closing must be kept into perspective since contact erosion during a closing operation is significantly less than during interruption. Unless the rate of change of the gap dielectric capability is extremely slow the pre-arcing time is bound to be considerably shorter than the interrupting arcing time. Furthermore it should be considered that due to the low instantaneous values of current at closing the energy input will be significantly lower than that which is seen during interruption.

The optimum switching angle for reducing, or eliminating prestrikes would be at voltage zero; nevertheless, this may present a problem because the maximum current peak is reached under these conditions due to the maximum asymmetry which is produced when the current flow, in an inductive circuit, is initiated at a voltage zero. As a consequence of the high current peak the electro-mechanical-stresses imposed on the circuit breaker are the highest. This translates into higher output energy requirements for the operating mechanism and in general larger structures for the circuit breaker. An alternate possibility for semi-synchronism would be to choose an optimum switching angle, one which in most cases would be between 30 and 45 electrical degrees.

Synchronous interruption. Theoretically a synchronous interrupter is one that changes instantaneously from a perfect conductor to a perfect isolator. This characteristic constitutes a physical impossibility for a mechanical device such as a conventional circuit breaker since, a finite gap would have to be developed in essentially zero time.

Nevertheless, it is possible to design a quasi-synchronous interrupter, one that separates its contacts consistently at a predetermined time just ahead of current zero. However, in order to achieve this a high operating contact speed is needed so that a contact gap, that is large enough to withstand the transient recovery voltage can be attained during the very short arcing period available. This approach has been successfully demonstrated in a number of experimental devices, for at least the last 30 years. A prototype design of a synchronous circuit breaker was installed and remained in service for over 15 years. A photograph of this prototype breaker is shown in figure 10.7.

In spite of the work that has been done and the knowledge that has been gained, there are still a number of practical problems associated with this concept. To reduce the mechanical stresses on the interrupter, due to the high operating forces which are required to accelerate the moving contacts it would be desirable to reduce their mass but, these lighter contacts, may no be able to carry the rated continuos current. To regain this capability a parallel set of primary contacts may need to be provided, but in doing so the operating scheme of the circuit breaker becomes more complex. If two sets of contacts, each moving at a different speed, are used then a complicated mechanical scheme, or two independent operating mechanisms would have to be provided for each pole assembly. A further complication arises, to a lesser extent with puffer SF_6 circuit breakers, but more definitely so with self pressure generating breakers that utilize the arc energy to generate the interrupting pressure, from the fact that by minimizing the arcing time there may not be sufficient energy and stroke for the former and time for the later to generate the required interrupting pressure.

Figure 10.7 ITE synchronous interrupting circuit breaker installed by American Electric Power.

The primary benefit to be derived from synchronous operation is a reduction in size and a decrease in the erosion of the arcing contacts. A relative comparison of the arc energy input for a non synchronous circuit breaker having a 12 millisecond arcing time and a synchronous interrupter with only a 1 millisecond arcing time is illustrated in figures 10.8 and 10.9 respectively.

It is unlikely that synchronous interruption will produce a noticeable improvement in the interrupting capacity because as it was shown in chapter 5 the recovery capability of an SF_6 interrupter is directly related to the rate of change of current at the instant of current interruption rather than to the magnitude of the current peak. With vacuum interrupters however some improvement may be expected because the duration of the coalescent arc mode may be significantly reduced and furthermore, depending upon the total magnitude of the current being interrupted the arc may remain in its diffuse mode for the full duration of the arcing time period.

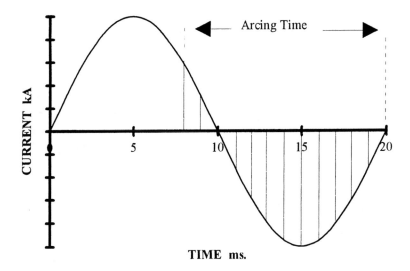

Figure 10.8 Arcing time for non-synchronous interruption.

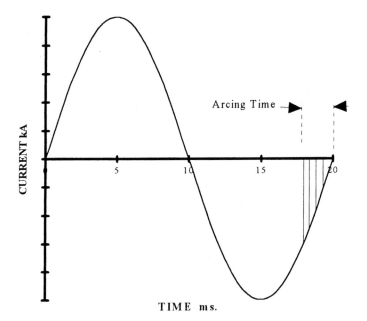

Figure 10.9 Maximum arcing time for a synchronous interruption.

For circuit breakers that have a characteristically long arcing time the opening speed tends to reach levels that are considered to be impractical. To get a feel for the opening velocities that are required consider for example a 72 kV circuit breaker where its contacts move at 3 meters per second and the minimum arcing time is 10 milliseconds, thus the minimum contact gap can be assumed as being approximately 30 millimeters. If arcing is to be limited to only 1 millisecond then the required opening speed is 30 meters per second. Considering the mass of the contacts and the linkages involved attaining this speed would be a very difficult task. Since circuit breaker design almost invariably entail compromises, some reduction on the contact erosion may be sacrificed to gain some reduction in the operating speed and the likely mechanical wear. The decision must be based upon sound evaluation of the technical and economic advantages of the concept talking into consideration the frequency of operations under fault conditions.

10.2 Condition Monitoring of Circuit Breakers

As it has been said before, circuit breakers constitute an important and critical component of the electric system, they are the last line of defense and consequently proper and reliable operation is paramount to the quality of power being delivered, to the promotion of customer satisfaction and most of all to the safety and integrity of the system [9].

To sustain the confidence level on this critical piece of equipment comprehensive maintenance programs have been established. These maintenance programs follow established standard guidelines and the recommendations of the manufacturer which generally are based on their operating experience. This practice may not only be inefficient but, it is also costly because of the down time required to perform these procedures. Additionally, it is not uncommon that problems develop following maintenance of otherwise satisfactorily performing equipment. A more logical approach may be to continually evaluate the condition of those components that through experience have been identified as being the most likely to fail and those whose failure could provoke a severe damage that would disrupt the service.

Historically, most of the of circuit breaker failures that have been observed in the field can be attributed to mechanical problems and difficulties related to the auxiliary control circuits. A number of studies, such as those made by CIGRE [10] provide with an excellent insight into the failure statistics of the components of a circuit breaker. The report indicates, as shown in figure 10.10, that 70% of the major failures in circuit breakers are of a mechanical nature, 19% are related to auxiliary and control circuits and 11% can be attributed to electric problems involving the interrupters or the current path of the circuit breaker.

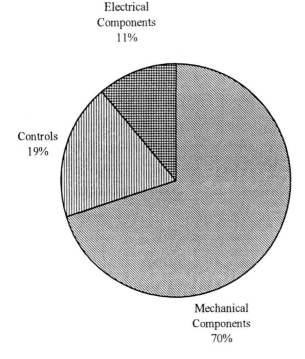

Figure 10.10 Types of circuit breaker failures by major components.

A further breakdown of the problems shows, (figure 10.11), that in the mechanical failure category approximately 16% involves compressors, pumps, motors, etc., 7% involve the energy storage elements, 10% is caused by control components, 7% are originated by actuators, shock absorbers, etc. and 3% result from failures of connecting rods and components of the power train.

In the group of electrical controls and auxiliary circuits, failure to respond to the trip and close commands accounts for 6% of the problems, 5% are due to faulty operation of auxiliary switches, 6% are caused by contactors, heaters etc. and 9% are attributed to deficiencies of the gas density monitors.

For those problems which are judged to be of electrical nature, the majority of them; 13% are due to failure of the insulation with respect to ground, 12% are due to the interrupters themselves and 2% are due to auxiliary interrupters or opening resistors or grading capacitors.

The just mentioned statistics can be used as a guideline for the selection of those components that should be monitored. Although the most desirable option would be to develop a system that constantly monitors critical components and

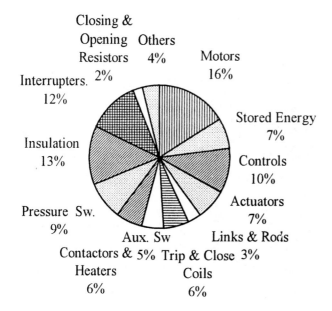

Figure 10.11 Percent of reported failures by components.

which is able to detect any deterioration that may occur over time and to predict, in a pro-active way, impending failures of mechanical components. This
task however, has proven to be rather elusive. A number of detection systems
have been investigated, including the use of acoustic signatures [11,12, 13],
but at least at this time these systems have not yet been translated into a viable
product and they still remain mostly in a laboratory environment. Furthermore
its reliability may be questioned primarily because of its complexity and its
high sensitivity, which makes it vulnerable to noise and to the influence of ex-
traneous sources. Simpler schemes may provide adequate protection, but naturally, a final choice should be based on an evaluation of the benefits against
the complexity and the difficulty of implementing the specifically required
monitoring function.

The information that is gathered by the monitoring system does not have to
be limited exclusively to evaluate the condition of the circuit breaker, but it
also may be used to enhance the accuracy of the controls for synchronous operation, if such operating option is available. It is entirely possible to use the
data to adjust the initiation of the closing or opening operation so as to compensate for variations in the making or breaking times that are due to the influence of the parameters that are being monitored.

10.2.1 Choice of Monitored Parameters

There are a significant number of parameters that can be chosen for monitoring, there are as well a variety of methods each having varying degrees of complexity for executing the monitoring function. The optimum system would be one that selects the most basic and important functions and thus minimizes the number of parameters that are to be monitored and yet it maximizes the effectiveness of the evaluation of the system that is being monitored.

In addition to optimizing the number of monitored parameters, the methods used to do the monitoring should be kept as simple and straight forward as possible. It will be desirable, if not essential, from the point of view of availability, cost and operating experience, that commercially available transducers, that are used in any related industry should be given preference and used where at all feasible.

10.2.1.1 Mechanical parameters

The most likely parameters to be monitored because of their significance and the simplicity of the monitoring scheme are given below. The list of likely candidates for monitoring and the possible methods that can be used are given only as a suggestion. Other parameters may be deemed to be as important and therefore they should be added to the list while others may be disregarded. The methodology may vary to suit the conditions of the application. What is important is to be able to have an indication, or warning if something is changing in the circuit breaker.

1. charging motors
2. contact travel distance and velocity
3. point of contact separation and contact touch
4. space heaters condition
5. trip coils and close coils
6. mechanism stored energy
7. breaker number of operations
8. ambient temperature

Charging motors. Encompassed under this denomination are all motors that are used for driving a gas compressor, a hydraulic pump, or for compressing a set of operating springs. From available statistics, motors, compressors and pumps exhibit the biggest share of failures.

A convenient way to monitor the condition of a motor would be to measure the starting and the running currents. These current values can then be used to calculate the torque of the motor, this torque value can then be used to judge the condition of the equipment or components that are being driven by the motor. For example, unusual increases in torque may indicate added friction which may be due to deteriorating bearings or galling of parts.

If the running time and the frequency of operation of the motor is compared to a baseline data established under standard operation it will be possible to detect leaks in a pneumatic or hydraulic system.

Contact travel and velocity. This could easily be considered as the most important function being monitored. It provides dynamic information about the operating components of the circuit breaker as a whole including not only mechanical links but also the interrupter contacts. The information that can be extracted from these measurements is always extremely valuable for judging the overall status of the circuit breaker and fortunately this is probably one of the simplest and easiest function to monitor.

From the measured travel characteristics by comparing the new data to a base line signature for the specific circuit breaker it should be possible to infer not only deterioration of linkages, but increased friction that could mean lack of proper lubrication and or deterioration of bearings for example.

For puffer circuit breakers the travel characteristics, when used in conjunction with the magnitude of the short circuit current being interrupted, would serve to indicate the degree of ablation of the interrupter's nozzle.

The contact, or breaker travel measurement can be easily consummated by monitoring the displacement or the rotation of the output shaft of the operating mechanism. The closing or opening velocities then can be obtained by finding the derivative with respect to time of the displacement measurement. This mathematical manipulation most likely would be done electronically with the assistance of a central processing unit where all the signals would be collected for evaluation and data storage.

The displacement measurement can be made using either contact or non-contact transducers. Sliding resistors, linear or rotating resistor potentiometers, step travel recorders, etc. are considered to be contact type transducers because there is an actual, physical connection between the transducer and the component being measured. Non-contact transducers are those such as optical motion sensors, proximity sensors, LTV sensors, etc. that do not require a physical connection between the sensor and the moving part. In all cases it is highly desirable that non-contact transducers be used to minimize errors that may be caused by the inherent inertia of the moving parts of a mechanical contact transducer.

It has been mentioned that it may be possible to compare the monitored instantaneous velocity to a pre-defined velocity for a particular circuit breaker operating under different sets of conditions, such as short circuit interruption, reduced energy from the mechanism, reduced ambient temperatures, etc. The evaluation of this comparison process would then be the criteria that is used to judge the condition of the circuit breaker.

Contact make and contact break. Contact break and contact make indications can be obtained either by direct, or indirect methods. If a direct indication is desired it can be obtained, when the measurement is made under load conditions by monitoring the voltage across the contacts. Additionally for a closing operation the measurement of the initiation of current flow provides an approximate indication of contact touch. The current indication can be used in conjunction with the no load travel measurement to compensate for normal prestriking.

For no load operations, or when indirect measurements are made, the methods that are used to determine contact displacement are applicable. The major draw back of this approach is that it fails to take into account the changes in the making or breaking of the contacts that may occur as a result of possible contact erosion.

Space heaters. Their function is simple and yet their failure may cause significant problems. Monitoring the integrity of the heater elements is a rather trivial task that can be done by simply circulating continuously a very small current. Another alternative is to use thermostats that are strategically located in close proximity to the heaters, one disadvantage of this method is that heaters are not energized all the time but only when the ambient temperature drops below a certain level and consequently a logic circuit that relates ambient temperature to heater temperature should be provided but still it does not provide continuous monitoring of the continuity of the heater element itself.

Trip and close coils. Monitoring of these components is a relatively simple task, although caution must be used to prevent the creation of parasitic current paths that could create mis-operations and problems with control relays.

In its most simple version the monitoring system would only require either a continuous or an intermittent high frequency signal to determine the electrical continuity of the coils.

Mechanism spring charge. The function referred to as spring charge is in reality a measurement of the kinetic energy that is stored in the operating mechanism. The measurement can be made by measuring the spring compression, in terms of its change of length, or the gas pressure of pneumatic accumulators used in conjunction with hydraulic mechanisms. The same type of measurements are also applicable for spring or pneumatic mechanisms.

The displacement measurement is made using the same type of instrumentation as that which is used for measuring the travel of the contacts of the circuit breaker. In general it is only necessary to detect the extreme limits of the displacement which corresponds to the limits for the required spring deflection that covers the specified range of operating forces.

Operations counter. This is a seemingly elementary piece of information and yet it is very significant specially on those breakers where the operating characteristics show variations that are related to the accumulated number of operations. This measurement is not something new and in most circuit breakers it has been routinely made if nothing else to keep a tally in order to perform maintenance at the next recommended maintenance interval.

Ambient temperature Monitoring the ambient temperature may also be considered as an elementary or trivial piece of information; but this information is needed to detect deviations from the historical operating characteristics of the circuit breaker under similar conditions to those being monitored. The measurement can also be useful to compensate for variations in the operating time in synchronous switching applications.

10.2.1.2 Electrical parameters

Dielectric failures and interrupter failures represent a high percentage of the listed reasons for circuit breaker problems. Although, many of these failures would take place without any prior warning, there are some cases where it would be possible to anticipate an impending failure based on some conditions which are generally well known and predictable as is the case with high levels of corona, high leakage currents, high moisture content and low insulating gas density . While some of these parameters can be monitored with only reasonable efforts, there are others which are difficult to monitor while the circuit breaker is energized and in service. What follows is again only a suggested list of significant electrical components that could be monitored.

Contact erosion and interrupter wear. Monitoring contact erosion and interrupter wear has a strong, direct influence upon the required maintenance frequency therefore, it is not only desirable, but beneficial to accurately evaluate the condition of the interrupters rather than to rely on the presently used method of simply adding the interrupted currents until the estimated accumulated duty that is given by applicable standards, or by the manufacturers recommendations, is reached.

Measurements of contact erosion, or interrupter wear can not be made directly, but it can be done conveniently by indirect methods using measurements of current and arcing time. The interrupted current can me measured using conventional instrumentation, such as current transformers, which are generally available in circuit breaker as an standard component. The arcing time, depending in the desired degree of sophistication, can be determined by optical detection of the arc, by measurement of the arc voltage, or simply by estimating the point of contact separation using the information given by the contact travel transducer and the duration of current flow from this time until it is interrupted.

The product of the current and the elapsed time from contact separation to current extinction gives a parameter to which the interrupter wear can be related. It is assumed that sufficient data has been collected during development tests relating contact erosion and nozzle ablation to ampere-seconds of arcing, and therefore by keeping track of the accumulated ampere seconds an adequate appraisal of the interrupter condition can be made.

Gas density. For SF_6 circuit breakers gas density rather than gas pressure is the parameter that should be monitored. To do this, it is possible to use commercially available temperature compensated pressure switches or alternatively, the density may be determined by electronically processing separate pressure and temperature signals. These signals can be combined by an algorithm representing the well known equation of state for the gas. Any deviation from a constant density line will indicate that there is a gas leak in the system and unless a massive catastrophic failure occurs, slow leaks can be alarmed and protective actions can then be implemented.

The selection of the corresponding constant density line depends on the initial filling conditions of the circuit breaker.

Gas moisture. Dielectric failures constitute one of the highest percentage mode of failure. A possible cause for these failures is a high moisture content in the insulating gas, which may lead to tracking along the surfaces of the insulating materials used in the interrupter assembly. To prevent moisture related problems monitoring the moisture content of the gas should be a high priority.

Checking for moisture content in a gas is an easier task when done during routine maintenance, when the circuit breaker is out of service. In most cases this routine inspection is all that should be needed because it is a standard practice to install inside of the interrupter moisture absorbing materials such as activated alumina. Furthermore, for a circuit breaker which has been properly dried and evacuated prior to filling, there is no reason for any significant amount of moisture to migrate inside of the interrupter unless there is a gas leak and the interrupter is then incorrectly refilled. Nevertheless, if it is desired to monitor the moisture content while the circuit breaker is in service, it is possible to do so using commercially available moisture monitoring instrumentation that can be readily connected to the gas system of any circuit breaker.

Partial discharges. The type and rate of deterioration of insulation and the time required for a breakdown of the insulation depends on the thickness of the material, on its chemical and thermal stability, on the applied stress and on the ambient temperature and humidity. Ultimate failure is usually caused by cumulative heating of the discharges, or by tracking across the material surfaces. Although there is no absolute basis for predicting the life of materials in the presence of electrical discharges, it has been demonstrated that insulation just can not be reliably operated above the discharge inception voltage. It is then desirable to detect the onset, as well as the magnitude of the discharges.

Measurement of these two characteristics is rather difficult even under carefully controlled conditions, such as in a test laboratory. The problem lies in the low magnitude of the signals involved, since the discharge pulses may be as low as one microvolt [14, 15].

An alternate solution would be to use acoustic or optical detectors. These methods can provide a reasonably accurate indication but mostly only as a qualitative evaluation of the discharges at or near the surface of the insulation.

Contact temperature. The temperature at or near the main contacts can be a good indicator for a number of possible potential problems with the circuit breaker. Large changes in contact temperature may be due to broken contact fingers, excessive burning of the main contacts, material degradation, oxide formation, weak contact springs, or even an improperly or not fully closed circuit breaker.

The temperature of the contacts, or the conducting parts can be measured using optical methods, or else it can be approximated by measuring the temperature of the surrounding gas, of the ambient temperature and of the continuous current that is being carried. Knowing the normal temperature rise of the breaker the corresponding temperature at these particular conditions can be calculated. The results then can be compared with what is expected from this circuit breaker based on previously obtained development data.

10.2.2 Monitored Signals Management

Considering the operating conditions under which the monitoring is done and the actions taken as a result of the information being obtained it would be possible to characterize these signals as being proactive or reactive.

Proactive signals would be those which do not depend on the circuit breaker being operated before the condition of a certain parameters could be determined. On the other hand reactive signals are those that can only be measured during a dynamic condition such as the opening or closing of the breaker contacts.

Proactive signals truly fulfill the intent of condition monitoring by giving the opportunity to take action either by sending an alarm or by preventing the operation of the circuit breaker when one of the components being monitored has failed. One thing that should be avoided is sending information continuously while all the conditions are normal. Decisions should be made at the breaker level as to what action must be taken.

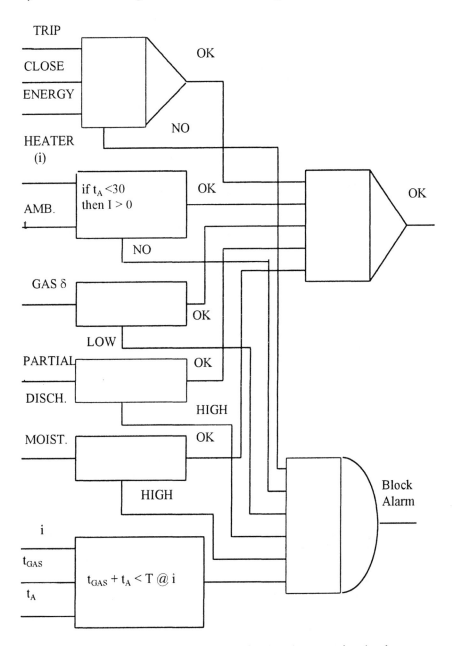

Figure 10.12 Logic diagram for action to be taken based on proactive signals.

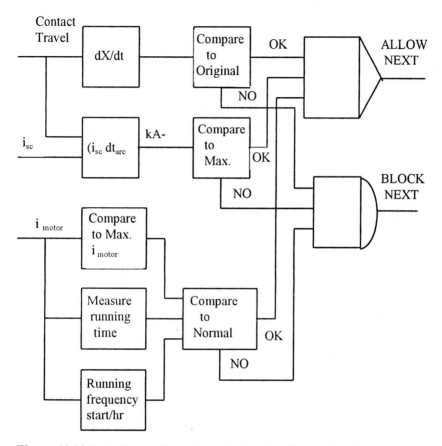

Figure 10.13 Logic diagram for action to be taken based on reactive signals.

To accomplish this objective it is possible to set up a switching logic scheme such as those shown in figures 10.11 and 10.12 for the proactive and the reactive signals respectively.

In the case of a reactive signal, if deviation from the maximum acceptable limits is observed, the action most likely to be taken would be to send an alarm accompanied by blocking of the next operation. The actual schemes will vary, depending upon the established local operating philosophies and consequently a number of widely diversified schemes may be found. As the technology further develops it may be possible to use neural networks and fuzzy logic to make more complex decisions once the networks have been properly trained and sufficient operational data is available that justifies the programmed actions.

REFERENCES

1. J. Beehler, L. D. McConnell, A New Synchronous Circuit Breaker for Machine Protection, IEEE Transactions T-PA&S 73: 668-672, March-April 1973.

2. L. D. McConnell, R. D. Garzon, The Development of a New Synchronous Circuit Breaker, IEEE Transactions T-PA&S 73: 673-681, March-April 1973.

3. R. D. Garzon, Synchronous Transmission Circuit Breaker Development, Final Report, Energy Research and Development Administration, CONS/2155-1, August 1976.

4. Controlled Switching A State Of The Art Survey Part II, Electra 164, TF 13.00.1: 39-69, Feb. 96.

5. Aftab Khan, D. S. Johnson, J. R. Meyer, K. B. Hapke, Development of a new synchronous closing circuit breaker for shunt capacitor bank energization, Sixty-First Annual International Conference of Doble Clients, Paper 5E, March 1994.

6. J. Roguski, J Experimental investigation of the dielectric recovery strength between the separating contacts of vacuum circuit breakers. IEEE Trans, Power Delivery 4. Vol 2: 1063-1069, 1989.

7. B. J. Ware, J. G. Reckleff,, G. Mauthe, G. Schett, Synchronous Switching of Power Systems, CIGRE Session 1990, Report No. 13-205.

8. G. Moraw, W. Richter, H. Hutegger, J. Wogerbauer, Point on Wave Controlled Switching of High Voltage Circuit-Breakers, CIGRE 13-02, 1988 Session.

9. H. Karrenbauer, C. Neumann, A Concept for Self-Checking and Autocontrol of HV Circuit Breakers and its Impact on Maintenance and Reliability, CIGRE Symposium, Berlin, Paper 120-03, April 1993.

10. Final Report of the Second International Inquire on High Voltage Circuit Breakers Failures and Defects in Service, Working Group 06 of study committee 13 (Reliability of High Voltage Circuit Breakers), June 1994.

11. L. E. Lundgaard, M. Runde, B. Skyberg, Acoustic Diagnosis of Gas Insulated Substations; a Theoretical and Experimental Basis, IEEE Trans. Power Delivery: 1751-1759, Oct. 1990.

12. V. Demjanenko, H. Naidu, A. Antur, M. K. Tangri, R. A. Valtin, D. P. Hess, S. Y. Park, M. Soumekh, A. Soom, D. M. Benenson, S. E. Wright, A Noninvasive Diagnostic Instrument for Power Circuit Breakers, IEEE Trans. Power Delivery: 656-663, April 1992.

13. D. P. Hess, S. Y. Park, M. K. Tangri, S. G. Vougioukas, A. Soom, V. Demjanenko, R. S. Acharya, D. M. Benenson, S. E. Wright, Noninvasive Condition Assessment and Event Timing for Power Circuit Breakers, IEEE Trans. Power Delivery: 353-360, January 1992.

14. T. Tanaka, M. Nagao, T. Okamoto, M. Ieda, H. Kärner, W. Kodoll, Electra No. 164, TF 15.06.01: 85-101, February 1996.

15. J. H. Mason, Discharge Detection and Measurements, Proc. IEE, Vol 112, No. 7, July 1965.

16. Application of Diagnostic Techniques for High Voltage Circuit Breakers, CIGRE 13.101, 1992.

APPENDIX: CONVERSION TABLES

TABLE I

CONVERSION TABLE FOR UNITS OF PRESSURE

	PASCAL (Pa) N/m^2	bar 0.1 MPa	psi lbs/in^2	atmospheres kg_{force}/cm^2	Torr 1/760 atm
1 Pa	1	10^{-5}	1.45×10^{-4}	10.2×10^{-6}	7.5×10^{-3}
1 bar	10^5	1	14.5	1.02	750
1 psi	6895	68.96×10^{-3}	1	70.27×10^{-3}	51.7
1 at	98100	0.981	14.23	1	736
1 Torr	133	1.33×10^{-3}	19.34×10^{-3}	1.36×10^{-3}	1

TABLE II

CONVERSION TABLE FOR UNITS OF

WORK ENERGY AND QUANTITY OF HEAT

	Joule (1 N . m = 1 Ws)	kWh	kcal	ft-lb
1 Joule	1	0.278×10^{-6}	0.239×10^{-3}	0.7376
1 kWh	3.6×10^{6}	1	860	0.3672×10^{6}
1 kcal	4187	1.1628×10^{-3}	1	427
1 ft-lb	1.3558	0.3766×10^{-6}	0.3238×10^{-3}	0.1383

TABLE III

CONVERSION TABLE FOR UNITS OF VOLUME

	cu. in.	cu. ft.	gallon	liter	cu. cm.
1 cu. in.	1	5.79×10^{-4}	4.33×10^{-3}	16.39×10^{-3}	16.39
1 cu. ft.r	1728	1	7.48	28.32	28.32×10^3
1 gallon	231	.134	1	3.79	3.79×10^3
1 liter	61	35.3×10^{-3}	264×10^{-3}	1	1×10^3
1 cu. cm.	.061	35.3×10^{-6}	264×10^{-6}	1×10^{-3}	1

TABLE IV

CONVERSION TABLE FOR UNITS OF

DENSITY

	lb/cu. ft	lb/gallon	g/cu. cm	kg/cu. m
1 lb/cu. ft.	1	.1337	.0160	16.02
1 lb/gallon	7.479	1	.1197	119.8
1 g/cu.cm	62.43	8.347	1	1000
1 kg/cu. m	.06243	8.347×10^{-3}	1×10^{-3}	1

TABLE V

CONVERSION TABLE FOR UNITS OF
THERMAL CONDUCTIVITY

	Btu/ft hr °R	cal/cm sec °K	joule / cm sec °K	watt/ cm °K	watt/ m °K
Btu/ft hr °R	1	4.134×10^{-3}	.0173	.0173	1.73
cal/cm sec °K	241.9	1	4.18	4.18	418.49
joule/cm sec °K	57.8	238.9×10^{-3}	1	1	100
watt/ cm °K	57.8	238.9×10^{-3}	1	1	100
watt/ m °K	.578	2.389×10^{-3}	.01	.01	1

INDEX